PROBABILITY SURFACE M
N INTRODUCTION WITH
AND FORTRAN PROGRAMMES

N. Wrigley

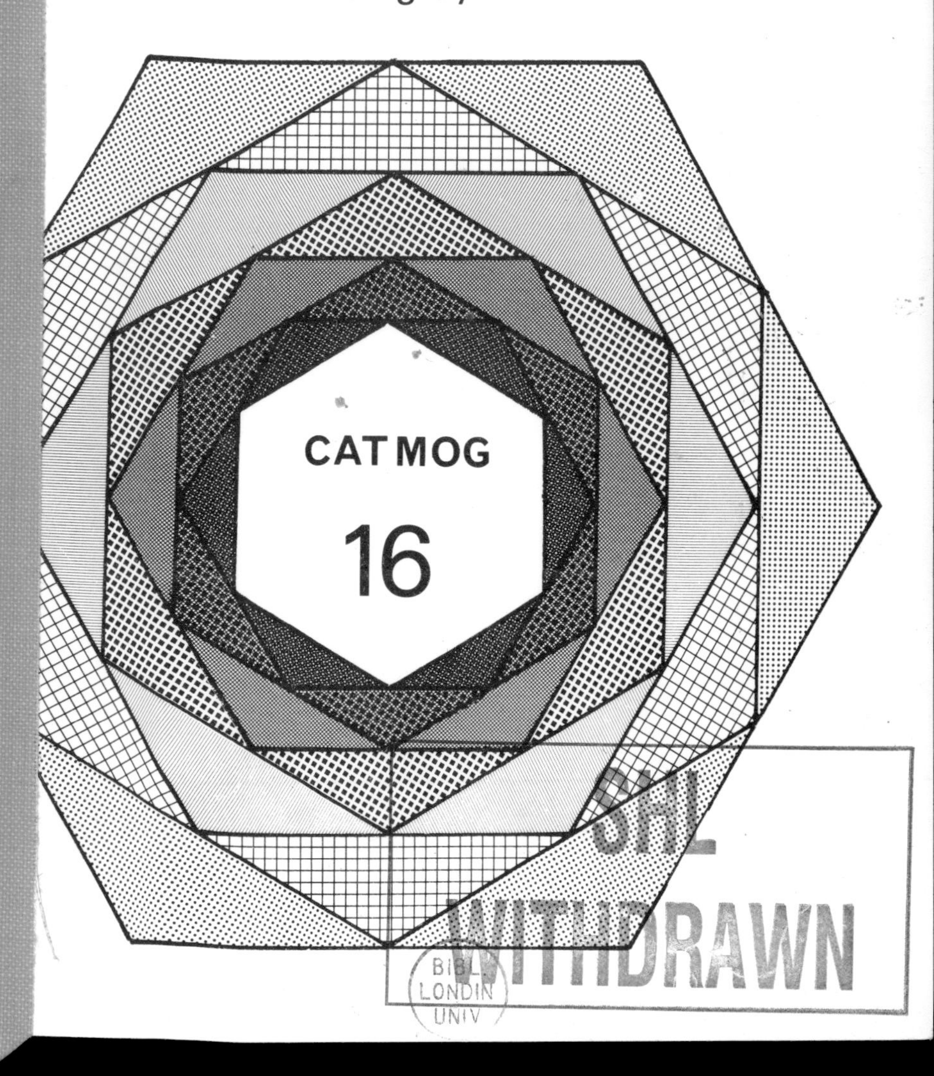

CATMOG

(Concepts and Techniques in Modern Geography)

CATMOG has been created to fill a teaching need in the field of quantitative methods in undergraduate geography courses. These texts are admirable guides for the teachers, yet cheap enough for student purchase as the basis of class work. Each book is written by an author currently working with the technique or concept he describes.

1. An introduction to Markov chain analysis - L. Collins
2. Distance decay in spatial interactions - P.J. Taylor
3. Understanding canonical correlation analysis - D. Clark
4. Some theoretical and applied aspects of spatial interaction shopping models - S. Openshaw
5. An introduction to trend surface analysis - D. Unwin
6. Classification in geography - R.J. Johnston
7. An introduction to factor analytical techniques - J.B. Goddard & A. Kirby
8. Principal components analysis - S. Daultrey
9. Causal inferences from dichotomous variables - N. Davidson
10. Introduction to the use of logit models in geography - N. Wrigley
11. Linear programming: elementary geographical applications of the transportation problem - A. Hay
12. An introduction to quadrat analysis - R.W. Thomas
13. An introduction to time-geography - N.J. Thrift
14. An introduction to graph theoretical methods in geography - K.J. Tinkler
15. Linear regression in geography - R. Ferguson
16. Probability surface mapping. An introduction with examples and Fortran programs - N. Wrigley
17. Sampling methods for geographical research - C. Dixon & B. Leach
18. Questionnaires and interviews in geographical research - C. Dixon & B. Leach

Other titles in preparation

*This series, Concepts and Techniques in Modern Geography is produced by the Study Group in Quantitative Methods, of the Institute of British Geographers.
For details of membership of the Study Group, write to the Institute of British Geographers, 1 Kensington Gore, London, S.W.7.
The series is published by Geo Abstracts, University of East Anglia, Norwich, NR4 7TJ, to whom all other enquiries should be addressed.*

£1-50

CONCEPTS AND TECHNIQUES IN MODERN GEOGRAPHY No 16

PROBABILITY SURFACE MAPPING

An introduction with examples and Fortran programs

by

Neil Wrigley
University of Bristol

CONTENTS

I acknowledge with thanks permission from AERE to use their Sub-routine VAO6AD from the Harwell Library.

I INTRODUCTION

(i) Purpose and pre-requisites

Since its introduction into geography in the early 1960's (Chorley and Haggett, 1965) trend surface mapping has become one of the most widely known and widely used methods of spatial analysis in geographical research. Basically the aim of trend surface mapping is to decompose or separate a spatial series into two components, a trend or regional component and an error or local component. This separation is accomplished by specifying and fitting an appropriate trend surface model; appropriate in the sense that it captures the underlying regional structure or trend of the spatial series and leaves a local component showing no discernible systematic spatial variation. A trend surface model is essentially a linear regression model in which the explanatory variables are the geographical co-ordinates of each site or locality in the spatial series.

During the past fifteen years, as geographers have become increasingly aware of the assumptions, limitations and potentialities of regression models, a deeper understanding of trend surface models has developed. However, despite this deeper understanding, the range of applicability of trend surface mapping has remained virtually unaltered. One of the major reasons for this is the fact that until recently the trend surface method has been unable to handle one of the most important types of data collected and analysed by geographers. There appears to have been universal acceptance that variables to be mapped using the trend surface method must be measured at a high level, at least interval scale. Unfortunately for the geographer, a large proportion of the variables measured and utilized in geographical research, particularly in human geography, are measured at lower levels, at the nominal or ordinal scales of measurement. That is to say they are categorizations of an unordered (nominal scale) or ordered (ordinal scale) nature. As a result many of the variables measured and utilized in the course of geographical research have traditionally been viewed as unmappable by the trend surface method, and this has severely limited its range of applicability. Recently an attempt has been made by the author to free the method from this limitation. A technique has been presented (Wrigley, 1977) which allows the extension of surface mapping to the realms of categorized data, and which as a result provides an approach suitable for the imperfect types of data which are faced in geographical research. The technique has been given the name probability surface mapping and the maps produced are termed probability surface maps.

In the original paper outlining the probability surface mapping method, there was only space to introduce and illustrate the method in the briefest of fashions. The aim of this monograph is to consider the topic at greater length, to present examples and computer programs, and give sufficient detail to make the method available to all those who require it for practical research purposes. In the following sections the method will be reviewed and worked examples and the necessary FORTRAN computer programs will be discussed. The monograph assumes the reader is familiar with trend surface mapping as it has traditionally been used in geography (see the earlier CATMOG 5 by Unwin, 1975a) and with multiple linear regression (see CATMOG 15 by Ferguson, 1977). Prior consideration of the original probability surface paper (Wrigley, 1977) and the author's CATMOG 10 (Wrigley, 1976) will prove useful but is not essential.

(ii) An introductory illustration

Before we embark upon discussion of the technical aspects of probability surface mapping the reader may find it helps to sustain his interest if he first has a taste of the potential of the method. Consider the following case. A marketing geographer wishes to investigate the trade area characteristics of a recently opened hypermarket. To assess the extent of the trade area and the level of market penetration within it, he conducts a household survey of shopping habits. Amongst other questions in the survey he elicits information from each housewife interviewed on whether or not she shops at the hypermarket and if she does, whether she shops there regularly or occasionally. Mapping the responses of the 144 housewives interviewed in the survey he draws the map shown in Figure 1. In this form the map is rather complex in appearance. There is some evidence of systematic spatial variation in the responses but the picture is by no means clear. By using the probability surface method, however, he is able to sift out the underlying systematic spatial variation in responses. He finds that the appropriate probability surfaces to fit are those shown in Figures 2, 3 and 4. These are, respectively, the probability surfaces of, regularly shopping at the hypermarket, occasionally shopping at the hypermarket and never shopping at the hypermarket. (The reader should note that these surface maps were produced automatically by the computer and the hand drawn contours were added to aid interpretation, using the key printed below each map).

Inspection of Figure 2 shows that the underlying regional structure of regularly shopping at the hypermarket is a 'dome' centred on the hypermarket. The probability of regularly shopping at the hypermarket falls away rapidly in all directions as one moves away from the hypermarket. In contrast, Figure 4 shows that the underlying regional structure of never shopping at the hypermarket is a flat-bottomed 'basin' centred on the hypermarket. The probability of never shopping at the hypermarket is low in the area surrounding the hypermarket but becomes high at the edges of the survey area. Finally, Figure 3 shows that the underlying regional structure of occasionally shopping at the hypermarket has a 'ring doughnut' type of structure, with the highest predicted probabilities 0.5 to 0.6 occurring in a broken ring or horseshoe some distance from the hypermarket. Immediately surrounding the hypermarket, the probability of occasionally shopping at the hypermarket falls to between 0.3 and 0.4, a level similar to that found in a ring some way from the hypermarket beyond the ring of higher probabilities.

Countless similar illustrations could have been provided. In the context of human geography for example, probability surfaces of aircraft noise annoyance around Manchester (Ringway) Airport have been described in Wrigley (1977), whilst in the context of physical geography and geology potential illustrations would include the use of probability surfaces of 'barren' or 'producing' oil wells in oil exploration, or probability surfaces of 'presence' or 'absence' of particular components in regional geochemical analyses.

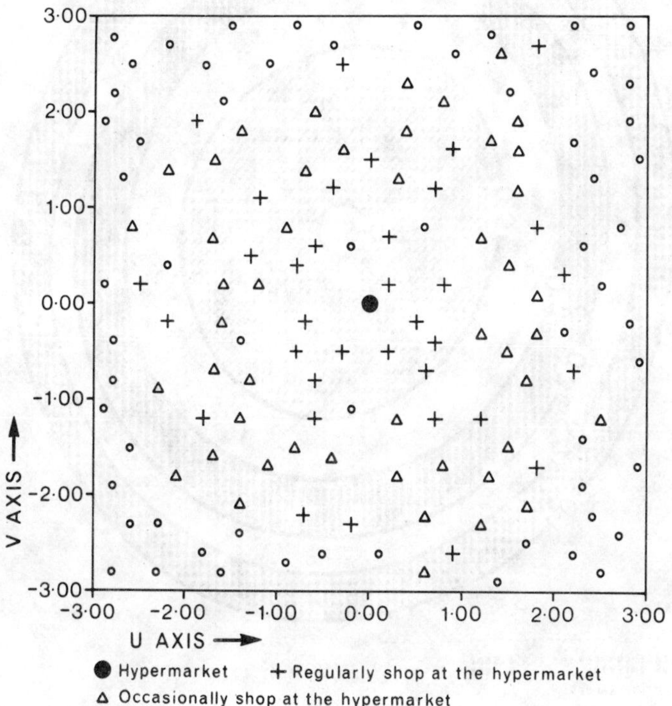

Fig. 1 Distribution of types of shoppers and non-shoppers around a hypermarket

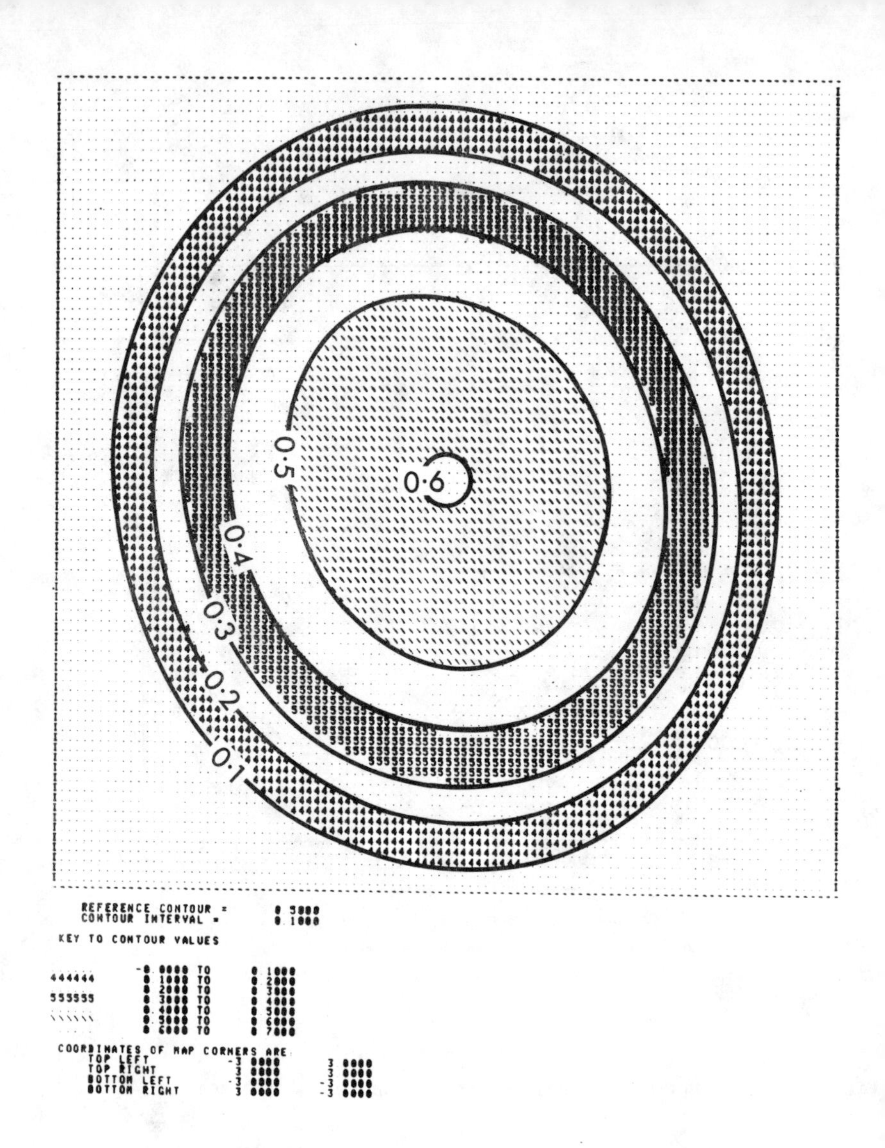

Fig. 2 2nd order probability surface of regularly shopping at the hypermarket

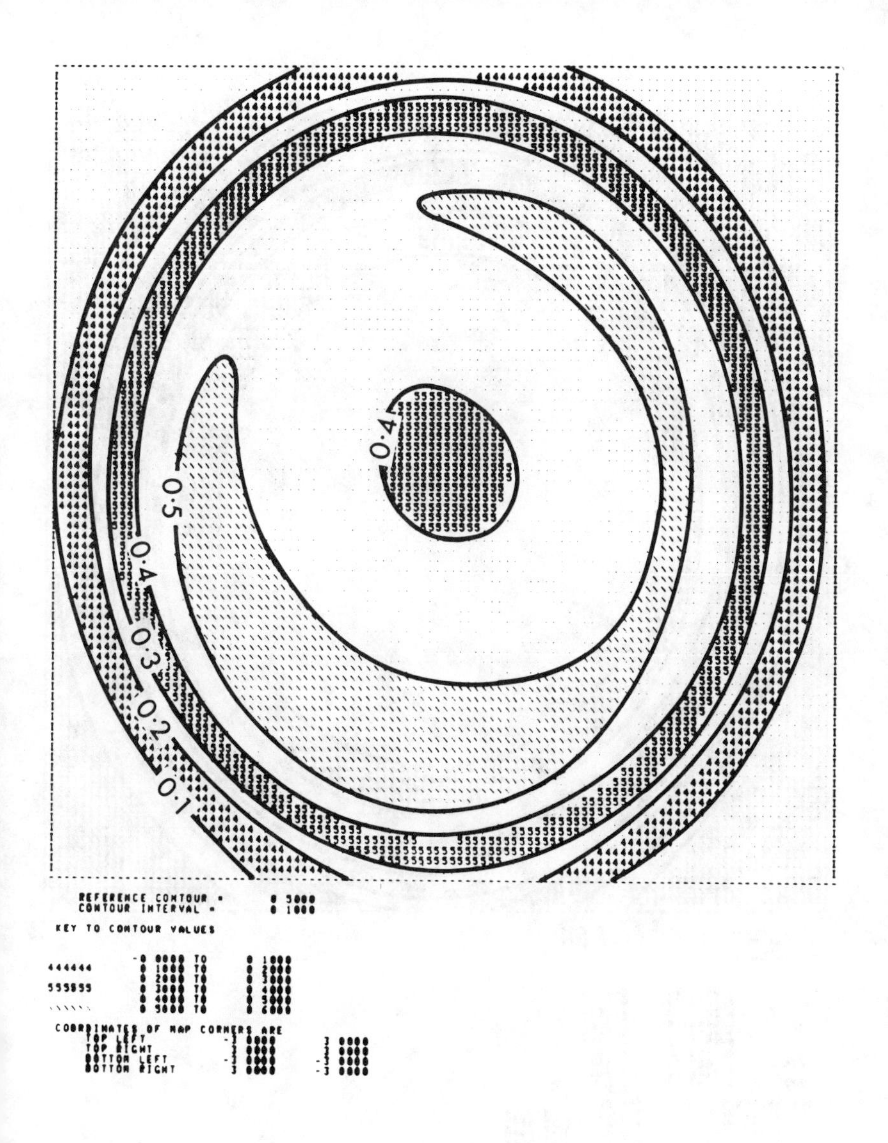

Fig. 3 2nd order probability surface of occasionally shopping at the hypermarket

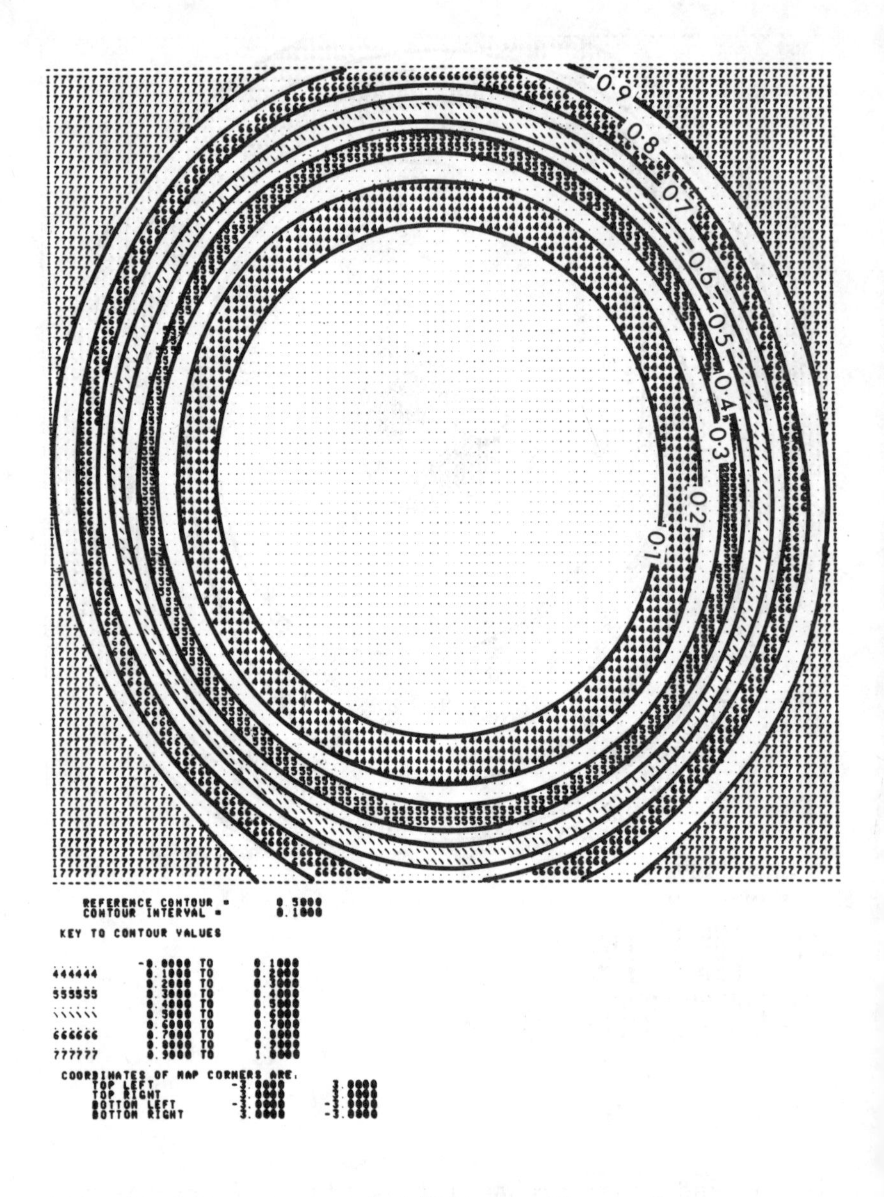

Fig. 4 2nd order probability surface of never shopping at the hypermarket

II THE TREND SURFACE MODEL AND THE SIMPLEST PROBABILITY SURFACE MODELS

(i) The form of the trend surface model

To understand probability surface models it is necessary to understand how they differ from traditional trend surface models. To help clarify these differences we will first briefly summarise the major features of trend surface models.

The traditional trend surface model can be written for each of a series of i localities (i=1,...N) as

$$Z_i = f(U_i, V_i) + \varepsilon_i \qquad (1)$$

Z_i is the response variable or variable to be mapped; a metrically (interval or ratio) scaled random variable, and U_i and V_i are the geographical co-ordinates of locality i. The trend surface model assumes that Z_i is composed of two components, a 'trend' or 'regional' component $f(U_i, V_i)$ and a 'local' component ε_i. In geographical research the most widely used form of the function $f(U_i, V_i)$ is the polynomial power series expansion which has the general form

$$f(U_i, V_i) = \alpha + \beta_1 U_i + \beta_2 V_i + \beta_3 U_i^2 + \beta_4 U_i V_i + \beta_5 V_i^2 + \ldots \qquad (2)$$

but other functional forms such as the double Fourier series (Bassett, 1972; Unwin, 1975a; Whitten, 1975) have also been used.

In the polynomial case, trend surface models of increasing complexity can be specified by including more and more terms in the power series expansion on the right hand side of equation (2). The more complex the trend surface model, the higher the so called 'order' of the trend or regional component of Z_i. The simplest polynomial trend surface model is of 1st order and has the form

$$Z_i = \alpha + \beta_1 U_i + \beta_2 V_i + \varepsilon_i \qquad (3)$$

The next most complex model is of 2nd order and has the form

$$Z_i = \alpha + \beta_1 U_i + \beta_2 V_i + \beta_3 U_i^2 + \beta_4 U_i V_i + \beta_5 V_i^2 + \varepsilon_i \qquad (4)$$

Three additional terms have been added to the 1st order model, and it can be seen that these new terms include squares or second powers of the co-ordinates. Higher order models follow the same pattern, the highest power included in the expansion at any point denotes the order of the trend surface model. For example a third order trend surface model has the form

$$Z_i = \alpha + \beta_1 U_i + \beta_2 V_i + \beta_3 U_i^2 + \beta_4 U_i V_i + \beta_5 V_i^2 + \beta_6 U_i^3$$
$$+ \beta_7 U_i^2 V_i + \beta_8 U_i V_i^2 + \beta_9 V_i^3 + \varepsilon_i \qquad (5)$$

In most of the following discussion, rather than distinguish a particular order of model, it will be useful to refer to polynomial trend surface models as a class. For this purpose we will use the general form of the expansion given in equation (2).

The trend surface model, equation (1), describes a stochastic dependence relationship between U_i, V_i and Z_i. The error term or stochastic disturbance ε_i is random and therefore for every pair of values U_i, V_i, there exists a whole probability distribution of possible values of Z_i. The trend surface model is therefore essentially a multiple regression model, and furthermore, because the forms of the function $f(U_i,V_i)$ used are normally linear in the unknown parameters (i.e. equation (2) is linear in the unknown parameters α, β_1, β_2...) the trend surface model is normally a multiple linear regression model. As a result, a full specification of the trend surface model must consist of the same assumptions as an equivalent multiple linear regression model (see Ferguson, 1977). In the case of most geographical applications, this means that the specification of the trend surface model must embody the same assumptions as in what is termed the classical normal multiple linear regression model. These assumptions are as follows.

(A1) The linearity assumption. The regression equation should be linear in the unknown parameters. For example, whereas equation (4) is nonlinear in the variables, it is linear in the parameters and thus is an intrinsically linear model.
(A2) The assumption that the values of the explanatory variables in the regression model can be measured without error.
(A3) The assumption that the values of the explanatory variables are fixed or nonstochastic. (This assumption can easily be relaxed.)
(A4) The assumption that no exact linear relationship exists between two or more of the explanatory variables. This is known as the no multicollinearity assumption.
(A5) The assumption that the number of observations exceeds the number of parameters to be estimated.
(A6) The zero error mean assumption. This is written $E(\varepsilon_i) = 0$. ($E(\varepsilon_i)$ is known as the expectation or expected value of ε_i, it is simply the mean of the probability distribution of possible values of ε_i.)
(A7) The constant error variance or homoscedasticity assumption. This is written $E(\varepsilon_i^2) = Var(\varepsilon_i) = \sigma^2$.
(A8) The independent error terms assumption. This is written $E(\varepsilon_i\varepsilon_j) = 0$, for $i \neq j$. In the case where i and j refer to geographical localities this assumption is known as the spatially independent error terms or no spatial autocorrelation assumption.
(A9) The assumption that ε_i is normally distributed.

In the case of the trend surface model, assumptions (A3) and (A6) imply that if we take expectations of both sides of (1), it follows that

$$E(Z_i) = E\left[f(U_i,V_i) + \varepsilon_i\right]$$
$$= f(U_i,V_i) \qquad (6)$$

(By assumption (A3), U_i and V_i are fixed values, thus $E(U_i) = U_i$ and $E(V_i) = V_i$. By assumption (A6) $E(\varepsilon_i) = 0$.) In the case of the widely used polynomial form of the function $f(U_i,V_i)$, equation (6) can be written as

$$E(Z_i) = \alpha + \beta_1U_i + \beta_2V_i + \beta_3U_i^2 + \beta_4U_iV_i + \beta_5V_i^2 \ldots \qquad (7)$$

This relationship gives the mean value of Z_i associated with the pair of geographical co-ordinates U_i and V_i for each of the i possible localities in a particular area. Across the i localities it defines what might be termed the 'true' or population trend surface. The parameters of this equation α, β_1, β_2 ... are unknown however, and consequently we must estimate their values from the observed Z_i, U_i, V_i values of a sample of the possible localities in the area. When the α and β parameters are estimated in this way we write them as $\hat{\alpha}$ and $\hat{\beta}$, and on the basis of their values (still using the polynomial form of the function) we define the equation

$$\hat{Z}_i = \hat{\alpha} + \hat{\beta}_1 U_i + \hat{\beta}_2 V_i + \hat{\beta}_3 U_i^2 + \hat{\beta}_4 U_i V_i + \hat{\beta}_5 V_i^2 \ldots \tag{8}$$

This gives $\hat{Z}_i$, the predicted or fitted value of the trend or regional component at locality i, which serves as a sample estimate of the 'true' or population trend component at locality i. Across the i localities equation (8) defines what might be termed the sample trend surface.

In practice few of the observed values of Z_i will lie exactly on the sample trend surface, most will lie either above or below it, and so the values of Z_i and $\hat{Z}_i$ will differ. This difference is called a residual and is designated e_i

$$Z_i - \hat{Z}_i = Z_i - (\hat{\alpha} + \hat{\beta}_1 U_i + \hat{\beta}_2 V_i + \hat{\beta}_3 U_i^2 + \hat{\beta}_4 U_i V_i + \hat{\beta}_5 V_i^2 \ldots)$$

$$= e_i \tag{9}$$

In general, because $\hat{\alpha}$, $\hat{\beta}_1$, $\hat{\beta}_2$ ... are likely to differ from the true values of α, β_1, β_2 ..., the residual e_i is different from the stochastic disturbance or error component term ε_i, for ε_i is given by the relationship

$$Z_i - E(Z_i) = Z_i - (\alpha + \beta_1 U_i + \beta_2 V_i + \beta_3 U_i^2 + \beta_4 U_i V_i + \beta_5 V_i^2 \ldots)$$

$$= \varepsilon_i \tag{10}$$

ε_i is a population term and cannot be observed. The value of the residual e_i can thus be regarded as a sample estimate of ε_i.

The method normally used by geographers to estimate the parameters of a trend surface model is the method of least squares. (See for example Unwin, 1975a, p.8-16, 19-21.) Under the assumptions (A1) to (A9) outlined above, the so called ordinary least squares (O.L.S.) estimators $\hat{\alpha}$, $\hat{\beta}_1$, $\hat{\beta}_2$ ... can be shown to be what are termed the best linear unbiased estimators (BLUES). Best linear unbiased estimators have a number of properties which intuitively we would like a 'good' estimator to possess. (See the discussion by Kmenta, 1971, p.154-93; Huang, 1970, p.26-32; Wonnacott and Wonnacott, 1970, p.40-47.) They are unbiased; in other words each estimator has a sampling distribution with a mean equal to the parameter to be estimated. Each estimator also has a variance which is smaller than that of any other unbiased estimator (best linear unbiasedness).

(ii) The simplest kind of probability surface model

Having summarised the major features of trend surface models we are now in a position to consider what happens if the response variable or variable to be mapped, Z_i, is not a metrically (interval or ratio) scaled random variable as it is assumed to be in traditional trend surface models, but is instead a categorized (nominal or ordinal scaled) variable.

The simplest case we can encounter of such a categorized response variable is a random variable with only two possible outcomes. For example, returning to the hypermarket trade area mapping illustration given earlier, the survey conducted might simply ask housewives to indicate whether they shop at the hypermarket or not, that is to say, to give a simple yes/no response. If we then code these two possible responses 1 and 0, 1 representing the response 'yes I shop at the hypermarket' and 0 representing the response 'no I do not shop at the hypermarket', and try to use such a response variable in the traditional trend surface model we will encounter three problems.

(i) The first problem we will encounter concerns the predicted values which are generated if we use the traditional trend surface model (1). Since Z_i in this case can only assume two different values, 1 and 0, $E(Z_i)$, the expected value of Z_i is a simple weighted average of the two possible values of Z_i with weights given by the respective probabilities of occurrence of the possible values. Purely arbitrarily we will say that the probability that $Z_i=1$ is P_i and that the probability that $Z_i=0$ is $1-P_i$. The expected value of Z_i is then

$$E(Z_i) = \{1 \times P_i\} + \{0 \times (1-P_i)\} = P_i \quad (11)$$

Using the result we found in equation (6) we then have

$$P_i = E(Z_i) = f(U_i,V_i) \quad (12)$$

In other words, it is useful and reasonable to interpret the expectation of of Z_i given the co-ordinates of locality i, U_i and V_i, as the probability of giving the specified response 'yes I shop at the hypermarket' at locality i. The problem with this interpretation however, concerns the predicted values $\hat{Z}_i$ generated using the traditional trend surface model (1). Using the polynomial form of the function, $\hat{Z}_i$ equals

$$\hat{Z}_i = \hat{\alpha} + \hat{\beta}_1 U_i + \hat{\beta}_2 V_i + \hat{\beta}_3 U_i^2 + \hat{\beta}_4 U_i V_i + \hat{\beta}_5 V_i^2 \ldots \quad (13)$$

In view of the probability interpretation of $E(Z_i)$, these predicted values are interpreted as predicted probabilities, i.e. $\hat{Z}_i = \hat{P}_i$. However, whereas probability is defined to lie between 0 and 1, the predictions (13) generated using the traditional trend surface model are unbounded and may take values from $-\infty$ to $+\infty$. Consequently the predictions may lie outside the meaningful range of probability and thus be inconsistent with the probability interpretation advanced.

(ii) The second problem we will encounter concerns the violation of the constant error variance assumption (A7). This follows from the fact that the error term

$$\varepsilon_i = Z_i - f(U_i,V_i) \quad (14)$$

can in this case only have one of two possible values

$$\varepsilon_i = \begin{cases} 1 - f(U_i,V_i) & \text{if } Z_i = 1 \\ - f(U_i,V_i) & \text{if } Z_i = 0 \end{cases} \quad (15)$$

These two possible values of ε_i must occur with probabilities P_i and $1-P_i$ respectively. Thus the assumption (A6) ($E(\varepsilon_i)=0$) implies

$$E(\varepsilon_i) = P_i\{1-f(U_i,V_i)\} + (1-P_i)\{-f(U_i,V_i)\} = 0 \qquad (16)$$

Solving for P_i or from (12) directly, we have that

$$P_i = f(U_i,V_i) \qquad (17)$$

$$1-P_i = 1 - f(U_i,V_i) \qquad (18)$$

The error variance can therefore be written

$$E(\varepsilon_i^2) = P_i\{1-f(U_i,V_i)\}^2 + (1-P_i)\{-f(U_i,V_i)\}^2$$

or using (17)

$$E(\varepsilon_i^2) = P_i(1-P_i)^2 + (1-P_i)(-P_i)^2$$

$$= P_i(1-P_i) \qquad (19)$$

Clearly the error variance is not a constant. Localities where P_i is close to 0 or close to 1 will have relatively low variances while localities where P_i is close to 0.5 will have higher variances. When the constant error variance assumption is violated the problem of heteroscedasticity is said to be present. Heteroscedasticity does not result in biased parameter estimates, but it does result in a loss of efficiency. In addition, heteroscedasticity implies that the estimated variances of the estimated parameters will be biased estimators of the true variances of the estimated parameters. If these biased estimators are used, then the statistical tests commonly used in trend surface mapping will be incorrect.

(iii) The third problem we will encounter, and an additional reason for not using the statistical tests employed in trend surface mapping is that in the categorized response variable case the error distribution is not normal. Assumption (A9) therefore does not hold and the commonly used statistical tests can not be applied since the tests depend on the normality of the errors.

As a means of handling a simple categorized response variable the traditional trend surface model is thus seriously deficient. The probability surface models which we will now consider attempt to overcome these deficiences.

We have seen that in the case of a simple categorized response variable with only two possible outcomes, the expected value of Z_i can be interpreted as a probability. That is to say, in our example $E(Z_i) = P_i$, where P_i is the probability of giving the response 'yes I shop at the hypermarket'. In view of this probability interpretation, the predicted values $\hat{Z}_i$ generated when the parameters of the trend surface model are estimated, are interpreted as predicted probabilities. In order to allow the probability interpretation it is necessary therefore that the condition

$$0 \leq \hat{P}_i \leq 1 \qquad (20)$$

is satisfied by the predicted values of the trend surface model. Unfortunately, as we have seen, this condition is not necessarily satisfied by the usual trend surface model for it produces predicted values which can range from $-\infty$ to $+\infty$. If we are to improve upon the traditional trend surface model as a means of handling categorized response variables we must therefore seek a model which produces predicted values which satisfy condition (20).

There are a number of potential models which do this, (see Wrigley, 1976, p.9-11; Domencich and McFadden, 1975, p.102-108; Pindyck and Rubinfeld, 1976, p.238-249) but perhaps the most convenient of them is based upon the logistic function.

$$P_i = \frac{e^{f(U_i,V_i)}}{1+e^{f(U_i,V_i)}} \tag{21}$$

As the value of $f(U_i,V_i)$ ranges from $-\infty$ to $+\infty$, P_i ranges in value from 0 to 1. As an alternative this model can be rewritten as follows to produce a linear model

$$1 + e^{f(U_i,V_i)} = \frac{e^{f(U_i,V_i)}}{P_i}$$

$$P_i = e^{f(U_i,V_i)} - e^{f(U_i,V_i)} P_i$$

$$P_i = (1-P_i)e^{f(U_i,V_i)}$$

$$\frac{P_i}{1-P_i} = e^{f(U_i,V_i)}$$

Then remembering that by the definition of a logarithm if $y=e^x$, $\log_e y=x$, we have

$$\log_e \frac{P_i}{1-P_i} = f(U_i,V_i) \tag{22}$$

The left hand side of this model is a transformation of P_i known as the <u>logit transformation</u> and we can abbreviate it as L_i. The important point to note about this transformation is that it increases from $-\infty$ to $+\infty$ as P_i increases from 0 to 1. What this means is that the predicted logit values $\hat{L}_i$ derived when the parameters of the model are estimated (using in this case the polynomial form of the function as an example)

$$\hat{L}_i = \log_e \frac{\hat{P}_i}{1-\hat{P}_i} = \hat{\alpha} + \hat{\beta}_1 U_i + \hat{\beta}_2 V_i + \hat{\beta}_3 U_i^2 + \hat{\beta}_4 U_i V_i + \hat{\beta}_5 V_i^2 \dots \tag{23}$$

can take any values in the range $-\infty$ to $+\infty$, but the predicted probabilities which can be found by substituting $\hat{\alpha}$, $\hat{\beta}_1$, $\hat{\beta}_2$ ... into equation (21) remain confined within the range 0 to 1.

We call the linear model (22) the linear logit probability surface model or simply the linear probability surface model. The nonlinear model (21) from which it was derived can therefore be termed the nonlinear probability surface model. Both satisfy the predicted probabilities condition (20) and both represent feasible alternatives to the traditional trend surface model for the categorized response variable case.

In the case of our example, P_i represents the probability of giving the response 'yes I shop at the hypermarket' at locality i. $1-P_i$ therefore represents the probability of giving the response 'no I do not shop at the hypermarket' at locality i, and the equivalent nonlinear model for this case is

$$1-P_i = \frac{1}{1+e^{f(U_i,V_i)}} \qquad (24)$$

It should be clear therefore that for the simplest categorized response variable, a dichotomous variable, one with only two possible outcomes, two probability surfaces can be mapped on the basis of models (21) and (24). In the case of our example, these two surfaces are the probability surface of shopping at the hypermarket, and the probability surface of not shopping at the hypermarket. It should also be clear that the probabilities of giving the two possible responses sum to one at each locality i. This means that in the two response category situation the nonlinear probability surface model is actually a set of two linked models (21) and (24); the models being linked by the fact that response probabilities must sum to one at each locality.

(iii) Estimating the parameters of the simplest probability surface models

Having found models which are more suited to the categorized response variable situation, how do we then estimate the parameters of these models? There are in fact two ways of doing this, a weighted least squares procedure and a direct maximum likelihood procedure.

In the case of the linear probability surface model we can use a least squares estimation method. Least squares is a method of estimation familiar to most geographers and the so called ordinary least squares (O.L.S.) method is the estimation procedure normally adopted for the traditional trend surface model. In the case of the linear probability surface model however, because the assumption (A7) of constant error variance is violated, the least squares method which must be employed is that known as weighted least squares (W.L.S). The method is discussed in the context of linear logit models in Wrigley (1976, p.12-18).

As an alternative to weighted least squares we can use a direct maximum likelihood procedure. This procedure uses the probability expressions (21) and (24) directly and thus it can be thought of as a method of estimating the parameters of the nonlinear probability surface model without first having to convert the model into the linear form (22).

Although both estimation methods give equivalent results in situations where the categorized variable is sampled at a large number of localities, and there are few parameters in the probability surface model to be estimated (i.e. in the polynomial case, when the order of the probability surface to be fitted is low), the maximum likelihood estimation method has significant practical advantages over the weighted least squares method. In the weighted least squares method it is necessary to group localities into sets (see Wrigley, 1976, p.12; Pindyck and Rubinfeld, 1976, p.249-250). This is a laborious procedure which does not lend itself to the development of the type of computer programs to be presented in Appendix 1. Moreover, it necessitates partitioning the continuous explanatory variables U_i and V_i (the geographical co-ordinates). In other words it involves reducing continuously distributed explanatory variables to a categorized form, with a consequent waste of information, and the risk of introducing bias into the estimation procedure. The only drawbacks to the maximum likelihood method are that it is less familiar to most geographers, and that it demands a numerical optimization computer routine. In this monograph the appropriate numerical optimization routine will be supplied and moreover it is hoped to show that in the case of probability surface models the maximum likelihood estimation procedure is readily understandable even by geographers totally unfamiliar with the method. As a result, from this point onwards, the method of probability surface mapping will be developed and illustrated assuming direct maximum likelihood estimation, and the use of the nonlinear form of the probability surface model.

In the case of the categorized response variable with only two possible outcomes, the first thing required when using the maximum likelihood estimation method is what is termed the likelihood of the set of observations. Because in this case the observations are drawn from a discrete distribution the likelihood is simply their joint probability of occurrence. This can be written

$$\text{Likelihood} = \Lambda = \prod_{i=1}^{N_1} P_i \prod_{i=N_1+1}^{N} (1-P_i) \tag{25}$$

(Π simply means multiply the elements together e.g. $\prod_{i=1}^{N} P_i = P_1 \times P_2 \times \ldots P_N$)

P_i is the probability that the housewife at locality i gives the response 'yes I shop at the hypermarket' and $1-P_i$ is the probability that she gives the response 'no I do not shop at the hypermarket'. N is the total sample size; N_1 is the total number of housewives in the sample who claim to shop at the hypermarket, and $N-N_1$ is the number who claim not to shop at the hypermarket. In other words, whenever $Z_i = 1$ (the housewife at locality i claims to shop at the hypermarket) the likelihood contains a term (21). Whenever $Z_i = 0$ (the housewife at locality i claims not to shop at the hypermarket) the likelihood contains a term (24). Substituting from (21) and (24) we thus have

$$\Lambda = \prod_{i=1}^{N_1} \frac{e^{f(U_i,V_i)}}{1+e^{f(U_i,V_i)}} \prod_{i=N_1+1}^{N} \frac{1}{1+e^{f(U_i,V_i)}} \tag{26}$$

As specified the likelihood depends upon a set of unknown parameters (e.g. in the case of the polynomial $f(U_i,V_i) = \alpha + \beta_1U_i + \beta_2V_i + \beta_3U_i^2 + \beta_4U_iV_i + \beta_5V_i^2 \ldots$). These parameters must then be estimated by taking as estimates the values which maximize the overall value of this likelihood equation. In practice, rather than maximize the likelihood itself, it is usual to maximize instead the logarithm of the likelihood. In the case of (26) this implies maximizing

$$\log_e \Lambda = \sum_{i=1}^{N_1} f(U_i,V_i) - \sum_{i=1}^{N} \log_e (1+e^{f(U_i,V_i)}) \qquad (27)$$

The maximum can be found by partially differentiating equation (27) with respect to its parameters and setting the partial derivatives equal to zero. The solution of the resulting set of equations yields the maximum likelihood parameter estimates. Provided no exact linear relationship exists between two or more of the explanatory variables, in other words, provided the data are not perfectly multicollinear, then the existence of a unique maximum is virtually certain in empirical samples of more than ten or twenty observations (Domencich and McFadden, 1975, p.111).

In practical terms the user of probability surface mapping need only know that the maximum likelihood estimation method employed has the logic described above. The computer routine described in Appendix 1(a) performs the estimation automatically.

(iv) Testing probability surfaces of progressively higher order

When discussing the widely used polynomial trend surface model (see II (i)) we noted that trend surface models of increasing complexity were denoted by the so called 'order' of the trend component they specified. A considerable amount of discussion in traditional polynomial trend surface mapping centres on statistical tests to determine what is the appropriate order of trend surface to fit; in other words what order of trend surface captures the underlying regional structure of the original spatial series (see Unwin, 1975a, p. 21-24). Now that the form of the simplest probability surface models and methods of parameter estimation have been discussed we must therefore consider the question of what is the appropriate order of probability surface to fit.

In trend surface mapping, for any surface of order N+1, two distinct hypotheses can be tested:

a) A null hypothesis of no trend. That is to say, we test whether all parameters associated with the explanatory variables U_i, V_i ... (note that this does not include the α parameter associated with the constant term) in the trend surface of order N+1 equal zero. This is a test of the significance of the surface in isolation from other surfaces.

b) A null hypothesis of no significant reduction in residual sum of squares, or increase in regression sum of squares, between a trend surface of order N and a surface of order N+1. This is a test of the improvement, if any, in the ability of a more complex trend surface of order N+1 to capture the underlying regional structure compared with a less complex surface of order N.

In the case of probability surface mapping, the same type of tests can be conducted, but instead of using residual or regression sums of squares and the F ratio, we now must use inferential tests based upon the maximized log likelihood value for different orders of probability surface. The test statistic we use (see Cox, 1970, p.88) is

$$\log_e \Lambda - \log_e \Lambda^* \tag{28}$$

$\log_e \Lambda$ is the maximized log likelihood of a set of unrestricted probability surface models, whereas $\log_e \Lambda^*$ is the maximized log likelihood of a set of restricted probability surface models embodying g constraints on the parameters of the set of unrestricted models. This test statistic is distributed asymptotically as one half chi-squared with g degrees of freedom.

Using this statistic, hypotheses equivalent to those in trend surface mapping, (a) and (b) above, can be tested.

a*) A null hypothesis of no trend, that is to say a test of whether all parameters associated with the explanatory variables U_i, V_i ... in the set of linked probability surface models of order N+1, equal zero. In this case $\log_e \Lambda$ is the maximized log likelihood for the set of linked probability surface models of order N+1, and $\log_e \Lambda^*$ is the maximized log likelihood for the set of linked probability surface models of order zero (in other words for the set of probability surface models containing only intercept/constant terms). For convenience we can write this as $\log_e \Lambda^*(C)$.

b*) A null hypothesis of no significant improvement in the ability of a linked set of more complex probability surface models of order N+1 to capture the underlying regional structure compared with a linked set of less complex probability surface models of order N. In this case $\log_e \Lambda$ is the maximized log likelihood for the set of probability surface models of order N+1, and $\log_e \Lambda^*$ is the maximized log likelihood for the set of probability surface models of order N.

In trend surface mapping, when determining the appropriate order of trend surface to fit it is wise to use in addition to the F ratio tests of hypotheses (a) and (b) described above, tests of the amount of systematic spatial variation in the maps of residuals from surfaces of different orders In recent years the ability to test residual maps for systematic spatial variation in this way has been significantly improved by the development of a spatial autocorrelation statistic for use with regression residuals by Cliff and Ord (1972) (see Cliff and Ord, 1973, p.122-127 for a trend surface mapping example).

In probability surface mapping, similar residual map tests would be a useful addition to the tests described in (a*) and (b*) above. To this end we first need therefore a definition of a residual. In the simple two response category case a residual can be defined as

$$Z_i - \hat{P}_i = e_i \tag{29}$$

Cox (1970, p.96) has suggested standardising such residuals to have a mean of zero and a unit variance as follows

$$\frac{Z_i - \hat{P}_i}{\sqrt{\hat{P}_i(1-\hat{P}_i)}} = e_i \tag{30}$$

Although this residual has useful properties, because Z_i can only take the value 0 or 1 its distribution like that of (29) will typically be highly non-normal. In particular, residuals close to the value zero will not occur except for extreme values of $\hat{P}_i$. In these locations where there is a high or low probability of the first response category being selected (or in terms of the example discussed above, of shopping at the hypermarket) the residuals have a very skew distribution. Thus in locations where the probability of selecting the first response category is high, the residuals are either small and positive or large and negative. In locations where the probability of selecting the first response category is low, the residuals are either small and negative or large and positive.

In attempts to extend the spatial autocorrelation statistic to the case of residuals (29-30) from probability surface maps, we are hampered not only by the extreme non-normality of the residuals but also by the fact that the Cliff-Ord spatial autocorrelation statistic for regression residuals is not applicable to residuals derived from a nonlinear model. Some form of generalisation of the Cliff-Ord regression residuals statistic is required. In the absence of such a statistic we might perhaps attempt purely visual or ad hoc statistical assessment of the presence of spatial autocorrelation amongst our mapped residuals, or employ in an informal manner Cliff and Ord's (1969) earlier spatial autocorrelation statistic for spatially distributed variables using the version of the test based upon the assumption of randomisation rather than that based upon the assumption of normality. These suggestions are however only tentative and much work remains to be done in this area.

(v) Goodness-of-fit statistics

In traditional trend surface mapping, so called 'goodness-of-fit' statistics are widely used to answer the question, how closely does a trend surface fit the original observations? (see Unwin, 1975a, p.14; Whitten, 1975, p.286-291). These statistics are based upon the R^2 value, the squared multiple correlation coefficient, and they range in value from either 0 to 1, or from 0% to 100%.

In probability surface mapping a goodness-of-fit statistic is available which fulfills the same purpose, but in this case it is based upon the likelihood ratio index (see Tardiff, 1976 and Domencich and McFadden, 1975, p.123). It has the form

$$\rho^2 = 1 - \frac{\log_e \Lambda}{\log_e \Lambda^*(C)} \tag{31}$$

As noted in II (iv), $\log_e \Lambda^*(C)$ is the maximized log likelihood for a set of linked probability surface models containing only intercept/constant terms (i.e. for the set of probability surface models of order zero). $\text{Log}_e \Lambda$ is the maximized log likelihood for the set of linked probability surface models which generate the probability surfaces whose goodness-of-fit is to be assessed. Like goodness-of-fit statistics in traditional trend surface mapping its value ranges from 0 to 1, or from 0% to 100% if multiplied by 100.

The reader should note however, that (31) is not the only goodness-of-fit statistic based upon the likelihood ratio which has been suggested. The economist John Cragg has suggested and used (see for example Baxter and Cragg, 1970, p.230) a 'pseudo' R^2 value which is defined as

$$R_p^2 = \frac{1-e^{\{2(\log_e\Lambda^*(C)-\log_e\Lambda)/N\}}}{1-e^{\{2\log_e\Lambda^*(C)/N\}}} \qquad (32)$$

The definition of its components is the same as in (31) with the addition of N which is the total number of observations in the sample. This statistic also ranges in value from 0 to 1 but it gives results which differ from those produced by (31). Clearly therefore the question of what constitutes an appropriate goodness-of-fit statistic is not yet fully resolved.

(vi) An empirical example

Now that we have discussed the form of the simplest probability surface models, methods of parameter estimation, and how to conduct inferential tests, we will work through an actual example of probability surface mapping. As in the discussion above, we will consider the case of a marketing geographer who wishes to investigate the trade area characteristics of a recently opened hypermarket. To assess the extent of the trade area and the level of market penetration within it he conducts a household survey of the shopping habits of 108 housewives in the area surrounding the hypermarket. Amongst other questions in the survey he asks each housewife whether she shops at the hypermarket or not. This information can either be portrayed in map form, as in Figure 5, or in tabular form, as in Table 1. Notice that in Table 1 the home of each housewife has been given a four figure map reference.

Given this survey data, the marketing geographer then wishes to fit probability surfaces of a polynomial type. To achieve this he uses the program supplied in Appendix 1(a). This program allows the user to fit, in a sequential manner, polynomial probability surfaces from order 1 up to order 4. For each surface order the program produces: probability surface maps for both the possible responses, i.e. the probability surface of shopping at the hypermarket and the probability surface of not shopping at the hypermarket; the maximized log likelihood; parameter estimates; standard errors; the variance-covariance matrix of the parameter estimates; predicted probabilities of giving each response for each of the sample respondents; raw residuals (see equation 29) and standardised residuals (see equation 30). Prior to this, the maximized log likelihood for the set of probability surface models containing only an intercept term is printed. This provides the $\log_e\Lambda^*(C)$ value required in the tests described in II (iv) and the goodness-of-fit statistics described in II (v). By using this program to fit probability surfaces up to order 3 he achieves the results shown in Table 2 and Figures 6, 7 and 8.

Table 2 gives the maximized log likelihoods for probability surface models of orders 0 to 3; the differences between these maximized log likelihoods; the differences expected under the null hypothesis (H_0) of no significant improvement between surface models of different orders (see II (iv) hypothesis b*), and the decision on the null hypothesis at conventional

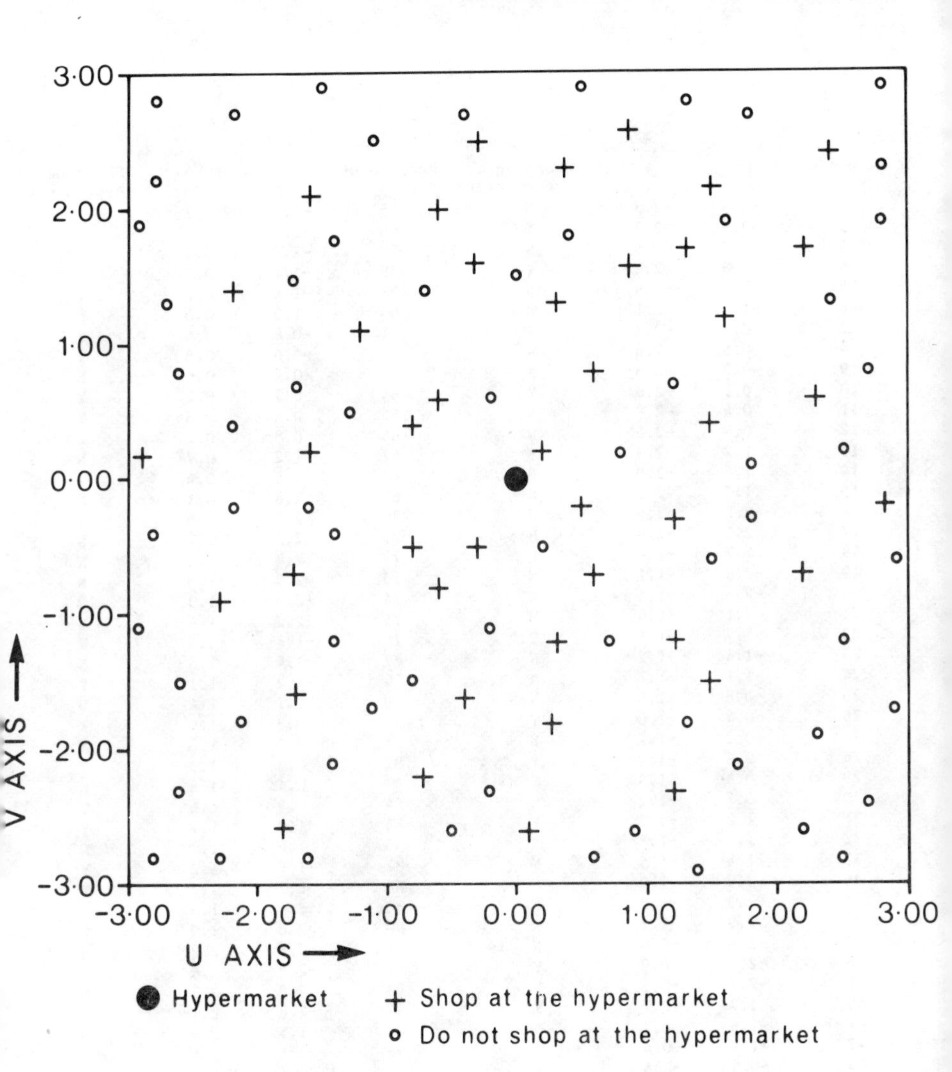

Fig. 5 Map plot of responses to shopping survey

Table 1

Respondent	U	V	Shopping at hypermarket
1	-2.8	-2.8	N
2	-2.6	-2.3	N
3	-2.3	-2.8	N
4	-1.8	-2.6	Y
5	-1.6	-2.8	N
6	-1.4	-2.1	N
7	-0.7	-2.2	Y
8	-0.5	-2.6	N
9	-0.2	-2.3	N
10	0.1	-2.6	Y
11	0.6	-2.8	N
12	0.9	-2.6	N
13	1.2	-2.3	Y
14	1.4	-2.9	N
15	1.7	-2.1	N
16	2.2	-2.6	N
17	2.5	-2.8	N
18	2.7	-2.4	N
19	-2.9	-1.1	N
20	-2.6	-1.5	N
21	-2.1	-1.8	N
22	-1.7	-1.6	Y
23	-1.4	-1.2	N
24	-1.1	-1.7	N
25	-0.8	-1.5	N
26	-0.4	-1.6	Y
27	-0.2	-1.1	N
28	0.3	-1.2	Y
29	0.3	-1.8	Y
30	0.7	-1.2	N
31	1.2	-1.2	Y
32	1.3	-1.8	N
33	1.5	-1.5	Y
34	2.3	-1.9	N
35	2.5	-1.2	N
36	2.9	-1.7	N
37	-2.8	-0.4	N
38	-2.3	-0.9	Y
39	-2.2	-0.2	N
40	-1.7	-0.7	Y
41	-1.6	-0.2	N
42	-1.4	-0.4	N
43	-0.8	-0.5	Y
44	-0.6	-0.8	Y
45	-0.3	-0.5	Y
46	0.2	-0.5	N
47	0.5	-0.2	Y
48	0.6	-0.7	Y
49	1.2	-0.3	Y
50	1.5	-0.5	N
51	1.8	-0.3	N
52	2.2	-0.7	Y
53	2.8	-0.2	Y
54	2.9	-0.6	N
55	-2.9	0.2	Y
56	-2.6	0.8	N
57	-2.2	0.4	N
58	-1.7	0.7	N
59	-1.6	0.2	Y
60	-1.3	0.5	N
61	-0.8	0.4	Y
62	-0.6	0.6	Y
63	-0.2	0.6	N
64	0.2	0.2	Y
65	0.6	0.8	Y
66	0.8	0.2	N
67	1.2	0.7	N
68	1.5	0.4	Y
69	1.8	0.1	N
70	2.3	0.6	Y
71	2.5	0.2	N
72	2.7	0.8	N
73	-2.9	1.9	N
74	-2.7	1.3	N
75	-2.2	1.4	Y
76	-1.7	1.5	N
77	-1.4	1.8	N
78	-1.2	1.1	Y
79	-0.7	1.4	N
80	-0.3	1.6	Y
81	0.0	1.5	N
82	0.3	1.3	Y
83	0.4	1.8	N
84	0.9	1.6	Y
85	1.3	1.7	Y
86	1.6	1.9	N
87	1.6	1.2	Y
88	2.2	1.7	Y
89	2.4	1.3	N
90	2.8	1.9	N
91	-2.8	2.2	N
92	-2.8	2.8	N
93	-2.2	2.7	N
94	-1.6	2.1	Y
95	-1.5	2.9	N
96	-1.1	2.5	N
97	-0.6	2.0	Y
98	-0.4	2.7	N
99	-0.3	2.5	Y
100	0.4	2.3	Y
101	0.5	2.9	N
102	0.9	2.6	Y
103	1.3	2.8	N
104	1.5	2.2	Y
105	1.8	2.7	N
106	2.4	2.4	Y
107	2.8	2.3	N
108	2.8	2.9	N

Table 2.

Order of surfaces	Maximized log likelihood	Difference between the maximized log likelihoods of surface orders: Observed	Expected under H_0: 1%	Expected under H_0: 5%	D.F.	Decision on H_0	Goodness of fit: ρ^2	Goodness of fit: R_p^2
0	−72.603							
		0.458	4.605	2.996	2	Fail to reject		
1	−72.145						0.0063	0.0114
		9.746	5.672	3.907	3	Reject at 1% level		
2	−62.399						0.1405	0.2329
		0.285	6.638	4.744	4	Fail to reject		
3	−62.114						0.1445	0.2388

significance levels. Also given are the values of the two goodness-of-fit statistics discussed in II (v).

Using test statistic (28) and the method explained in II (iv) hypothesis b*, the first time a step from surface order N to surface order N+1 is encountered that does not significantly improve our ability to capture the underlying regional structure, we should stop fitting surfaces of higher order and accept the probability surface models of order N. There is, however, a generally acknowledged exception to this rule in the trend surface context (see Whitten, 1975, p. 289) which also applies in the case of the example under consideration. The exception occurs where the underlying regional structure is a symmetrical 'dome' or 'basin'. In this case, first order surfaces will capture little of this underlying structure and the improvement over zero order surfaces will be insignificant. The second order surfaces, however, will capture a significant amount of this underlying regional structure, and it is clearly worthwhile extending the models to this higher order.

Table 2 shows clearly that it is the second order probability surfaces which are the appropriate surfaces to fit. The increased complexity (i.e. the increased number of parameters to be estimated) associated with moving from second order probability surface models to third order models is associated with only a minor improvement in the maximized log likelihood, and consequently there is no advantage to be gained.

The conclusions drawn using Table 2 are borne out in an examination of the probability surfaces themselves. Figures 6 (a) and (b), 7 (a) and (b), and 8 (a) and (b), show respectively the first, second and third order probability surfaces of shopping at the hypermarket and not shopping at the hypermarket. Clearly the third order surfaces add little to the picture of the underlying regional structure given in Figures 7 (a) and (b) by the second order surfaces. This underlying regional structure is a 'dome' centred on the hypermarket in the case of Figure 7 (a), the probability surface of shopping at the hypermarket, and a 'basin' centred on the

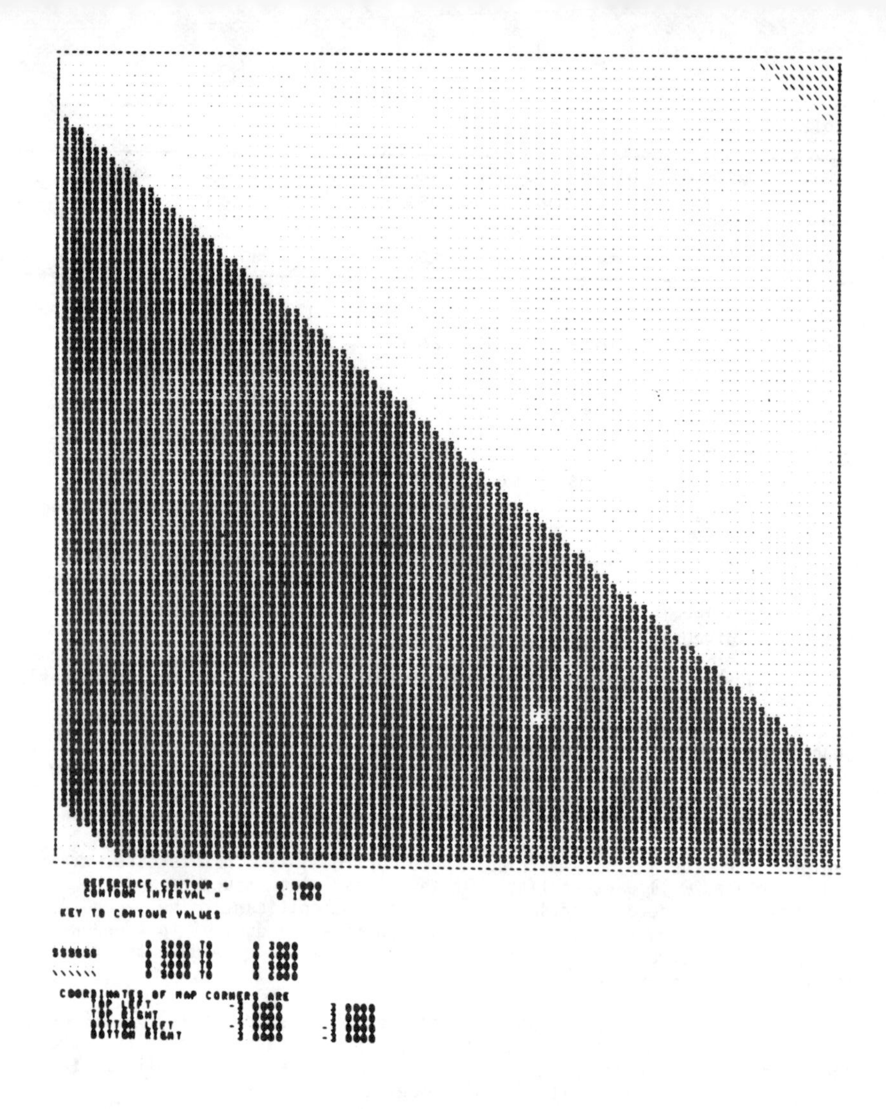

Fig. 6(a) 1st order probability surface of shopping at the hypermarket

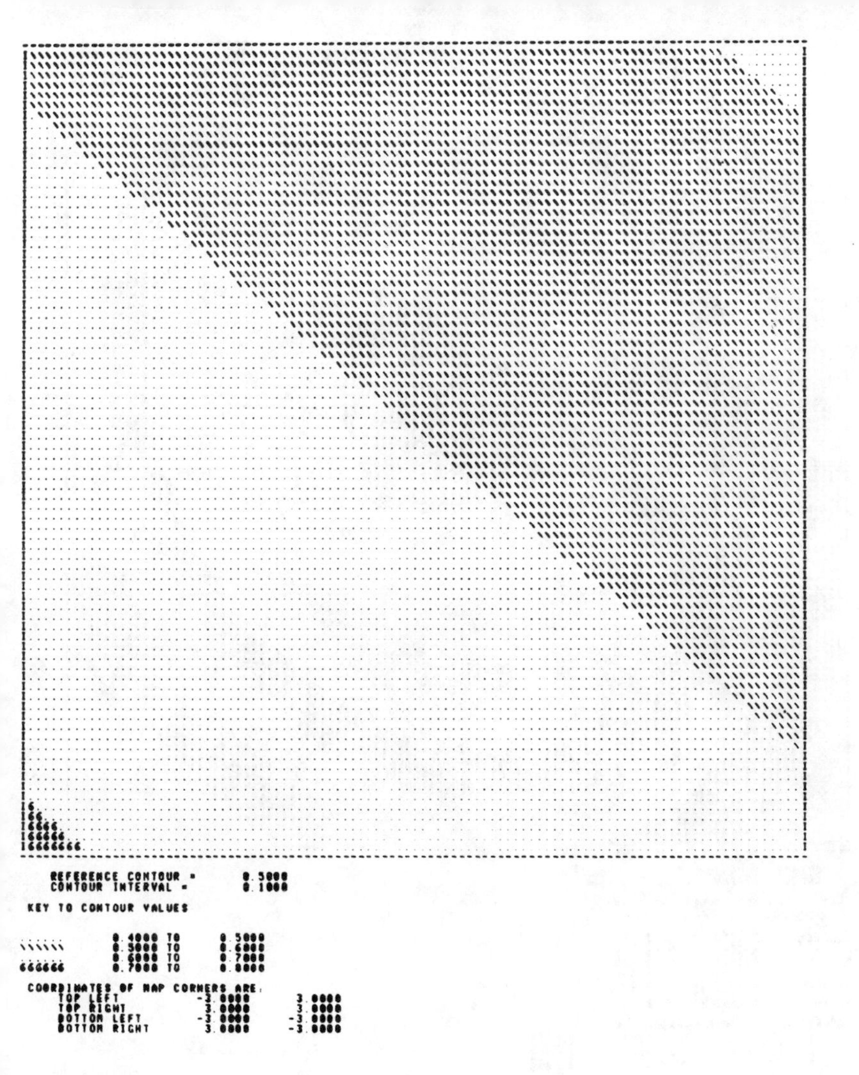

Fig. 6(b) 1st order probability surface of not shopping at the hypermarket

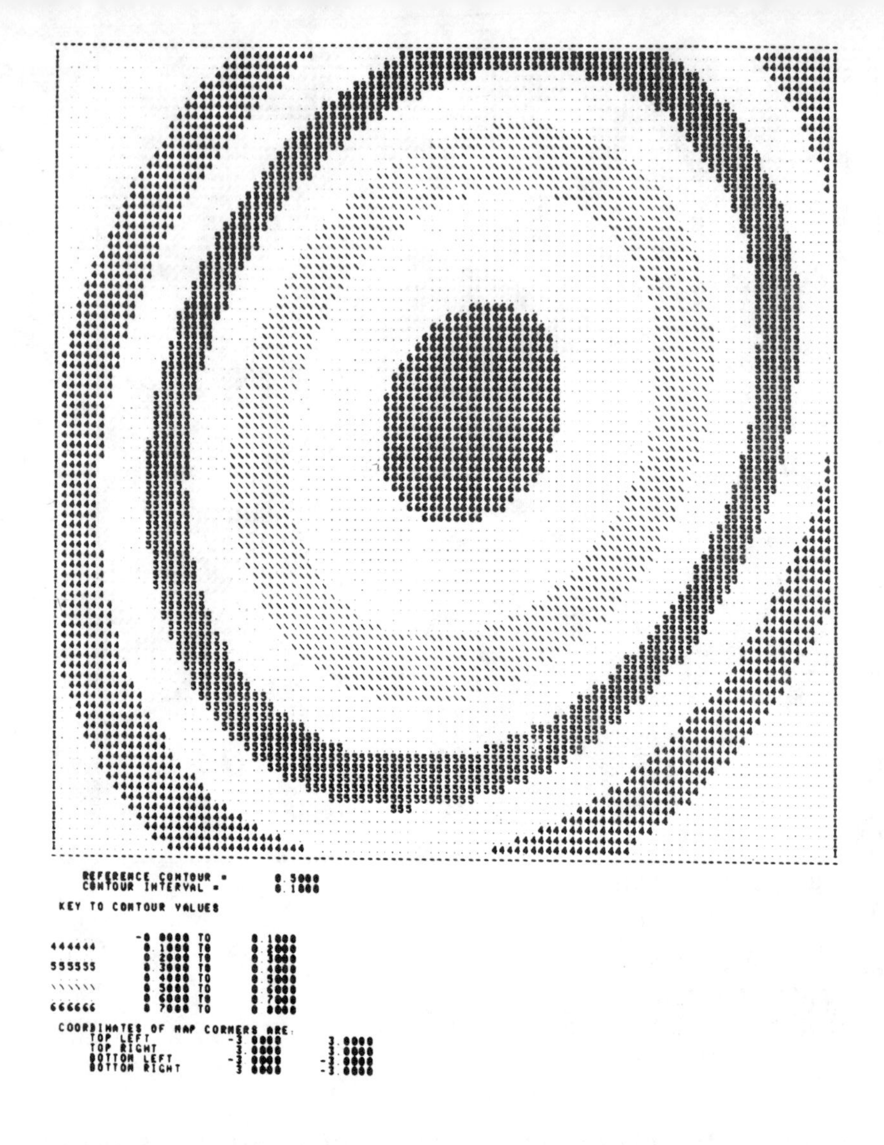

Fig. 7(a) 2nd order probability surface of shopping at the hypermarket

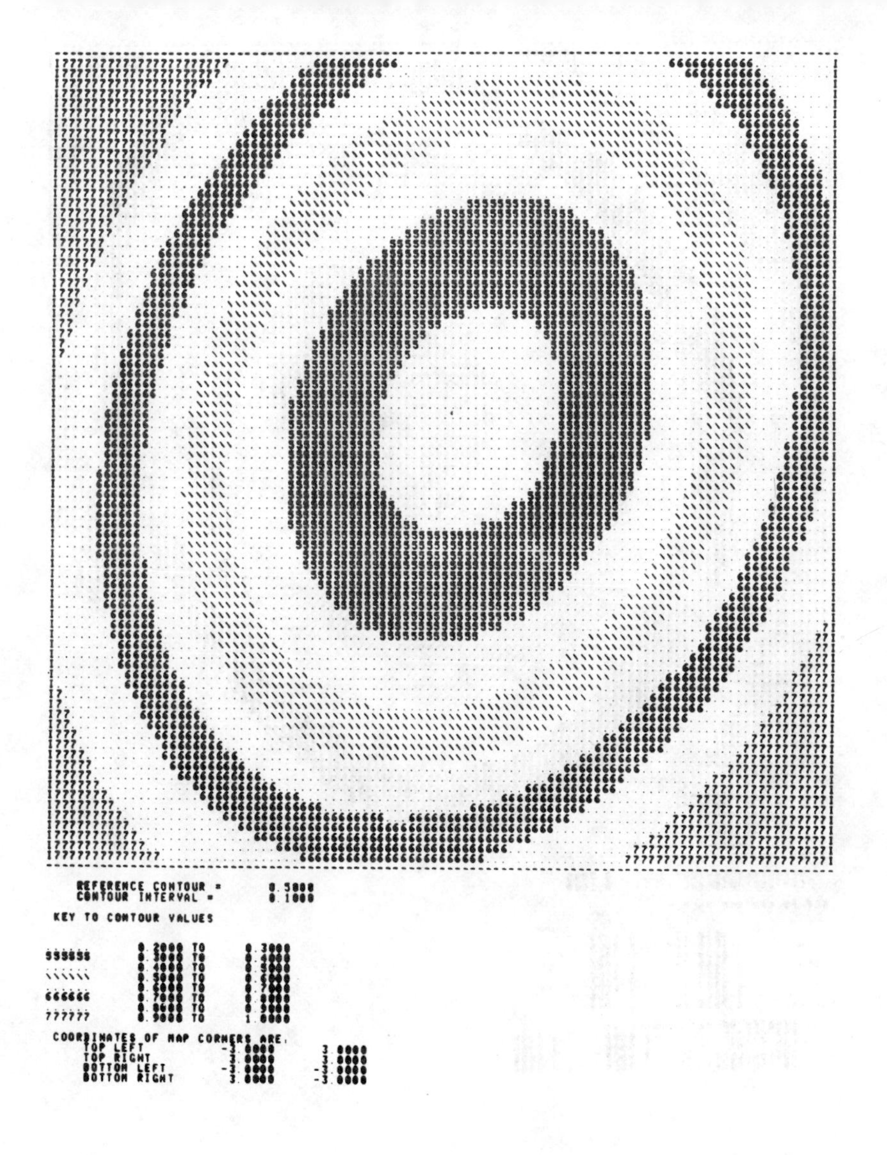

Fig. 7(b) 2nd order probability surface of not shopping at the hypermarket

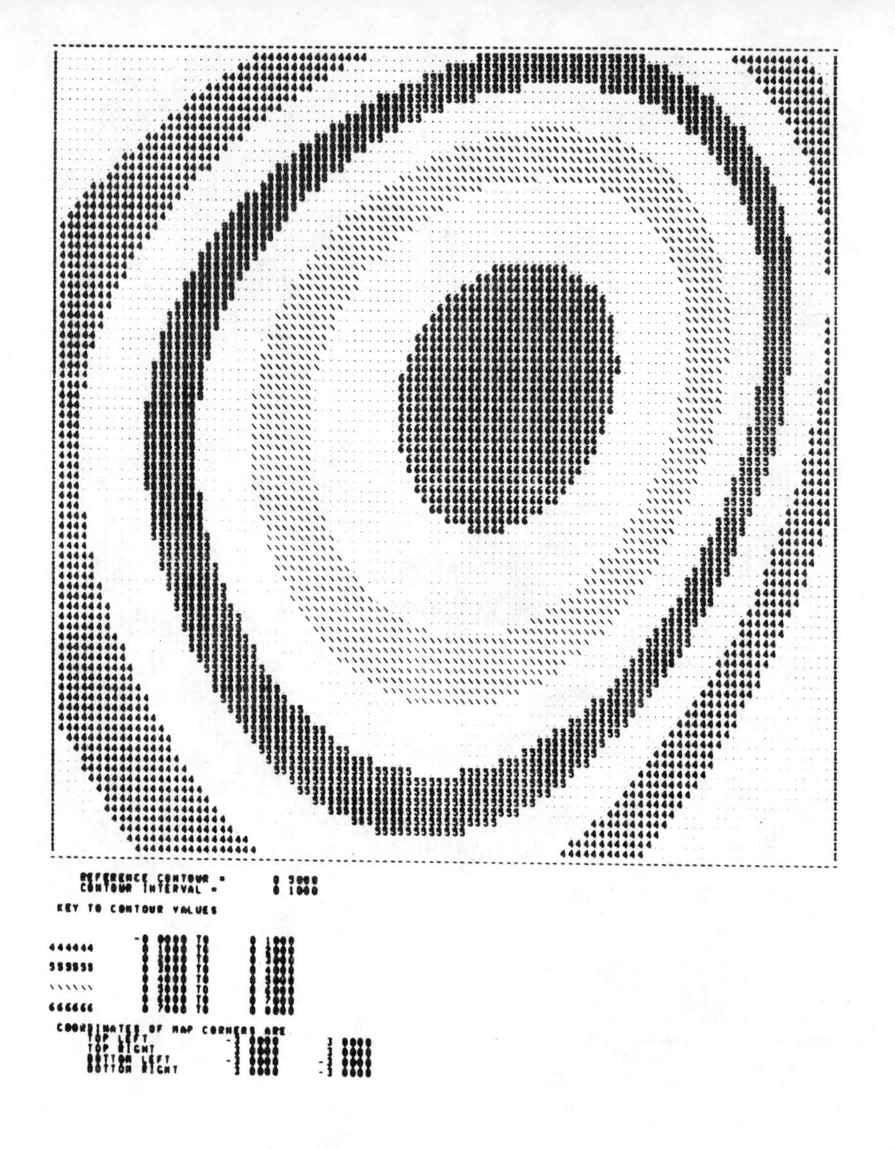

Fig. 8(a) 3rd order probability surface of shopping at the hypermarket

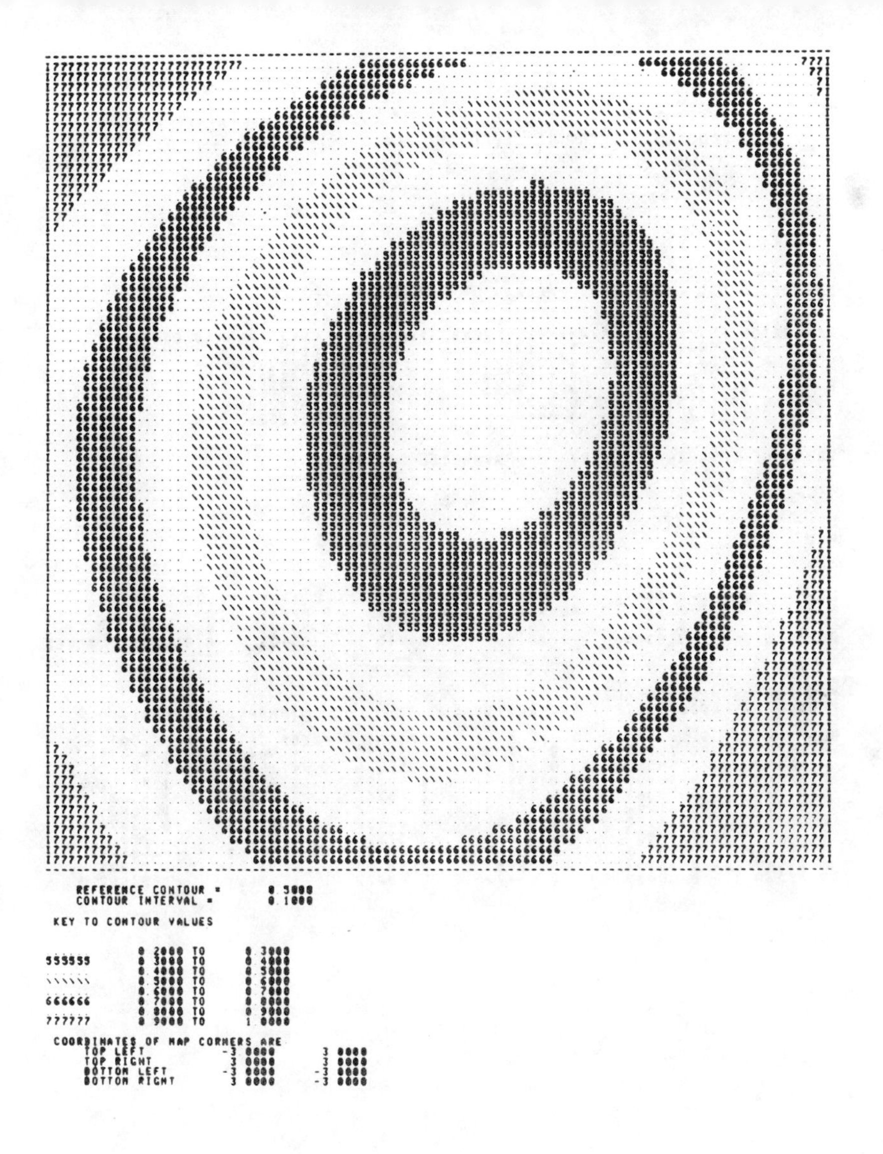

Fig. 8(b) 3rd order probability surface of not shopping at the hypermarket

Table 3 Part of the information printed by the probability surface mapping program for each surface order fitted

MAXIMIZED LOG LIKELIHOOD VALUE = -62.39922

PARAMETER ESTIMATES	STANDARD ERRORS	PARAMETER EST/STANDARD ERROR
0.9685544	0.3718716	2.6045401
0.0925432	0.1335535	0.6929300
0.1037769	0.1289748	0.8046292
-0.2911291	0.0917605	-3.1727065
0.1056475	0.0876147	1.2058200
-0.2151272	0.0840652	-2.5590529

VARIANCE-COVARIANCE MATRIX OF PARAMETER ESTIMATES

COLUMN	1	2	3	4	5	6
ROW						
ROW 1	0.13829					
ROW 2	0.00732	0.01784				
ROW 3	-0.00434	-0.00170	0.01663			
ROW 4	-0.02452	-0.00085	-0.00022	0.00842		
ROW 5	0.00967	-0.00161	0.00036	-0.00129	0.00768	
ROW 6	-0.01896	-0.00109	-0.00013	0.00129	-0.00223	0.007

FOR EACH SAMPLE LOCALITY

PREDICTED PROBABILITES		RAW RESIDUALS		STANDARDISED RESIDUALS	
RESP CAT 1	RESP CAT 2	RESP CAT 1	RESP CAT 2	RESP CAT 1	RESP CAT 2
0.20247	0.79753	0.79753	-0.79753	1.98469	-1.98470
0.41442	0.58558	0.58558	-0.58558	1.18869	-1.18869
0.31504	0.68496	0.68496	-0.68496	1.47451	-1.47451
0.26739	0.73261	0.73261	-0.73261	1.65526	-1.65526
0.38710	0.61290	0.61290	-0.61290	1.25831	-1.25831
0.55868	0.44132	0.44132	-0.44132	0.88878	-0.88878
0.62193	0.37807	0.37807	-0.37807	0.77967	-0.77967
0.50731	0.49269	0.49269	-0.49269	0.98548	-0.98548
0.51847	0.48153	0.48153	-0.48153	0.96371	-0.96371
0.39530	0.60470	0.60470	-0.60470	1.23681	-1.23682

hypermarket in the case of Figure 7 (b), the probability surface of not shopping at the hypermarket.

The probability surface maps shown in Figures 6, 7 and 8 are the line printer maps produced by the computer. Contours occur at the dividing line between symbols. (As an exercise the reader may wish to draw in these contours following the example of Figures 2, 3 and 4.) A key is automatically printed below each map to aid interpretation. The small page size of this monograph makes this key difficult to read but the reader should remember that in practice the computer prints these maps and the key at a much larger size.

In addition to the inferential tests presented in Table 2, residuals from the probability surface maps should be mapped and an attempt made to assess the amount of systematic spatial variation in these maps (see II (iv)). As can be seen in Table 3 (which shows part of the information printed by the program for each surface order fitted) residuals are automatically printed by the program presented in Appendix 1(a) and construction of these residual maps will be left as an exercise for the reader.

III EXTENDING THE SIMPLEST PROBABILITY SURFACE MODELS

(i) The generalised probability surface model

The case we have examined so far, the case in which the categorized response variable has only two possible outcomes, is only the simplest probability surface problem. There are far more cases in which the geographer is faced with categorized response variables with more than two possible outcomes. We call these polychotomous rather than dichotomous variables. Probability surface models can, however, be readily extended to handle such cases. For example, in our hypermarket survey instead of housewives simply indicating whether they shop at the hypermarket or not, let us presume that they indicate whether they regularly shop at the hypermarket, occasionally shop at the hypermarket, or never shop at the hypermarket. In other words, let us return to the introductory illustration in I (ii). For this three response category case, the nonlinear probability surface model generalises to the form

$$P_{r/i} = \frac{e^{f_r(U_i,V_i)}}{\sum_{s=0}^{2} e^{f_s(U_i,V_i)}} \qquad r = 0,1,2 \tag{33}$$

$P_{r/i}$ represents the probability of the rth response being given by the housewife at locality i, and the 3 responses 'regularly shop at the hypermarket', 'occasionally shop at the hypermarket' and 'never shop at the hypermarket' are arbitrarily coded 0,1 and 2. It should be stressed that there is no assumption of ordering in this coding.

In the polynomial case, equation (33) therefore can be written

$$P_{r/i} = \frac{e^{\alpha_r+\beta_{1r}U_i+\beta_{2r}V_i+\beta_{3r}U_i^2+\beta_{4r}U_iV_i+\beta_{5r}V_i^2 \cdots}}{\sum_{s=0}^{2} e^{\alpha_s+\beta_{1s}U_i+\beta_{2s}V_i+\beta_{3s}U_i^2+\beta_{4s}U_iV_i+\beta_{5s}V_i^2 \cdots}} \quad r=0,1,2 \qquad (34)$$

As was implicit in the simple 2 response category model, one of the response categories is now arbitrarily chosen and the parameters associated with that category are set to zero. The reasons for this are given by Wrigley (1976, p.22-24), and Pindyck and Rubinfeld (1976, p.256-259). If we choose the response category 'never shop at the hypermarket' and set its associated parameters to zero in equation (34), this implies

$$\alpha_2 = \beta_{12} = \beta_{22} = \beta_{32} = \beta_{42} = \beta_{52} \ldots = 0 \qquad (35)$$

Remembering that by definition $e^0 = 1$, the generalised probability surface model (34) thus becomes

$$P_{r/i} = \frac{e^{\alpha_r+\beta_{1r}U_i+\beta_{2r}V_i+\beta_{3r}U_i^2+\beta_{4r}U_iV_i+\beta_{5r}V_i^2 \cdots}}{1+\sum_{s=0}^{1} e^{\alpha_s+\beta_{1s}U_i+\beta_{2s}V_i+\beta_{3s}U_i^2+\beta_{4s}U_iV_i+\beta_{5s}V_i \cdots}} \quad r=0,1,2 \qquad (36)$$

Using for convenience the general function $f(U_i,V_i)$ instead of the polynomial, we can write the generalised probability surface model (36) out as

$$P_{0/i} = \frac{e^{f_0(U_i,V_i)}}{1+\sum_{s=0}^{1} e^{f_s(U_i,V_i)}}$$

$$P_{1/i} = \frac{e^{f_1(U_i,V_i)}}{1+\sum_{s=0}^{1} e^{f_s(U_i,V_i)}} \qquad (37)$$

$$P_{2/i} = \frac{1}{1+\sum_{s=0}^{1} e^{f_s(U_i,V_i)}}$$

This shows that the generalised probability surface model for the three category situation is actually a set of three linked models; just as in the two category situation we saw that there was a set of two linked models given by equations (21) and (24). In (37) the reader should note that the numerator of the right hand side of the equation for $P_{2/i}$ is 1 because

$$e^{f_2(U_i,V_i)} = e^{\alpha_2+\beta_{12}U_i+\beta_{22}V_i+\beta_{32}U_i^2+\beta_{42}U_iV_i+\beta_{52}V_i^2 \ldots}$$

$$= e^0 = 1 \qquad (38)$$

This extension of the nonlinear probability surface model to the three response category case, generalises to any number of categories. For example, with J possible response categories the model (33) simply takes the form

$$P_{r/i} = \frac{e^{f_r(U_i,V_i)}}{\sum_{s=0}^{J-1} e^{f_s(U_i,V_i)}} \qquad r = 0, \ldots J-1 \qquad (39)$$

and this can be expanded out, just as (37) was expanded from (33), to show that it is actually a set of J linked models.

(ii) <u>Estimating the parameters of the extended probability surface models</u>

Maximum likelihood estimation of the parameters of the 3 response category model demands a simple expansion of the likelihood for the two response category case. Instead of (25) we now have

$$\Lambda = \prod_{i=1}^{N_1} P_{0/i} \prod_{i=N_1+1}^{N_2} P_{1/i} \prod_{i=N_2+1}^{N} P_{2/i} \qquad (40)$$

or substituting from equation (37)

$$\Lambda = \prod_{i=1}^{N_1} \frac{e^{f_0(U_i,V_i)}}{1+\sum_{s=0}^{1} e^{f_s(U_i,V_i)}} \prod_{i=N_1+1}^{N_2} \frac{e^{f_1(U_i,V_i)}}{1+\sum_{s=0}^{1} e^{f_s(U_i,V_i)}} \prod_{i=N_2+1}^{N} \frac{1}{1+\sum_{s=0}^{1} e^{f_s(U_i,V_i)}} \qquad (41)$$

N is the total sample size; N_1 is the number of housewives in the sample who claim to shop regularly at the hypermarket; N_2-N_1 is the number of housewives who claim to shop occasionally at the hypermarket, and $N-N_2$ is the number of housewives who claim never to shop at the hypermarket.

As in the simple two response category case, instead of maximizing the likelihood itself, it is usual to maximize the logarithm of the likelihood. In the 3 response category case the log likelihood function (27) generalises to

$$\log_e\Lambda = \sum_{i=1}^{N_1} f_0(U_i,V_i) + \sum_{i=N_1+1}^{N_2} f_1(U_i,V_i) - \sum_{i=1}^{N} \log_e\left(1+\sum_{s=0}^{1} e^{f_s(U_i,V_i)}\right) \qquad (42)$$

The maximum can be found by partially differentiating equation (42) with respect to its parameters and setting the partial derivatives equal to zero. The solution of the resulting set of equations yields the maximum likelihood parameter estimates. The computer routine described in Appendix 1(b) performs the estimation automatically.

Extension of the maximum likelihood estimation procedure to the case of a 4 response category model or ultimately to a J response category model follows exactly the same principles as those outlined above in the extension of the estimation procedure from the 2 response category case to the 3 response category case. The computer routine described in Appendix 1(c) performs the estimation automatically for the 4 response category case.

(iii) Testing probability surfaces and goodness-of-fit statistics

The inferential tests and goodness-of-fit statistics discussed in II (iv) and II (v) apply equally as well in the multiple response category case as in the two response category case. The reader should therefore consult II (iv) and II (v). The only difference is that in the multiple response category case the definitions of a residual (29) and (30) must be generalised to the form

$$Z_{r/i} - \hat{P}_{r/i} = e_{r/i} \qquad (43)$$

$$\frac{Z_{r/i} - \hat{P}_{r/i}}{\sqrt{\hat{P}_{r/i}(1-\hat{P}_{r/i})}} = e_{r/i} \qquad (44)$$

$\hat{P}_{r/i}$ is the predicted probability that the individual at locality i will choose the rth response category, and $Z_{r/i} = 1$ if category r is chosen and 0 otherwise. In the multiple response category case there are thus as many residual maps as there are response categories.

(iv) An empirical example

To illustrate the use of probability surface mapping in the multiple response category case, we will use once again the example of a marketing geographer who conducts a survey to investigate the trade area characteristics of a recently opened hypermarket. In this case he conducts a survey of the shopping habits of 144 housewives in the area surrounding the hypermarket and amongst other questions in the survey he asks each housewife whether she regularly shops at the hypermarket, occasionally shops at the hypermarket, or never shops at the hypermarket. Once again this information can be portrayed in map form as in Figure 9 (which is Figure 1 repeated for convenience), or in tabular form as in Table 4. Given this survey data, he then fits probability surfaces of a polynomial type using the program supplied in Appendix 1(b) and he achieves the results shown in Table 5.

Table 5 gives the maximized log likelihoods for probability surface models of orders 0 to 3; the differences between these maximized log likelihoods; the differences expected under the null hypothesis of no significant improvement between surface models of different orders; and the decision on the null hypothesis at conventional significance levels. Also given are the values of the two goodness-of-fit statistics discussed in II (v).

As in the example of Section II, the fact that the underlying regional structure of each response is symmetrical, results in the first order surfaces capturing little of this underlying structure, and consequently

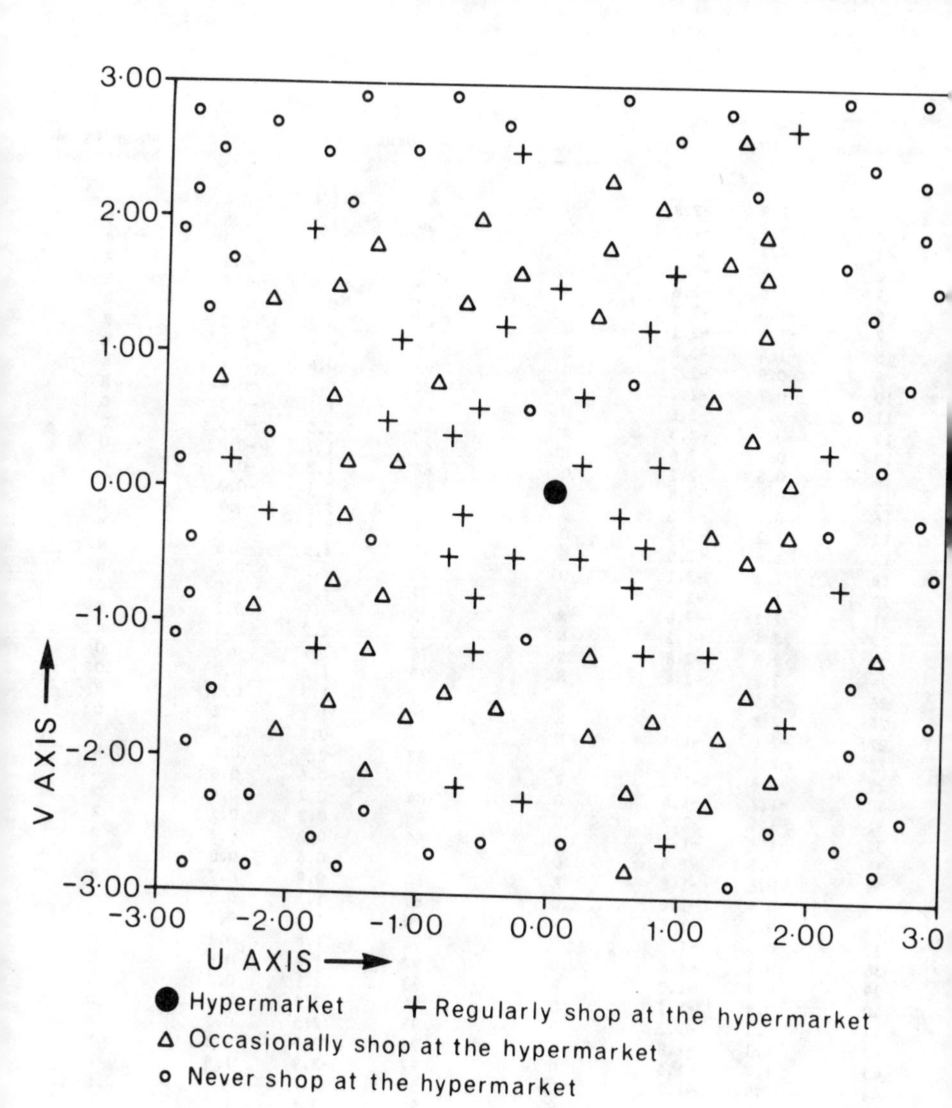
3·00
2·00
1·00
0·00
-1·00
-2·00
-3·00
V AXIS
-3·00
-2·00
-1·00
0·00
1·00
2·00
3·0
U AXIS
Hypermarket
Regularly shop at the hypermarket
Occasionally shop at the hypermarket
Never shop at the hypermarket

Fig. 9

Table 4

Respondent	U	V	Shopping at hypermarket	Respondent	U	V	Shopping at hypermarket
1	-2.8	-2.8	N	55	-1.4	-0.4	N
2	-2.6	-2.3	N	56	-1.3	-0.8	O
3	-2.3	-2.8	N	57	-0.8	-0.5	R
4	-2.3	-2.3	N	58	-0.7	-0.2	R
5	-1.8	-2.6	N	59	-0.6	-0.8	R
6	-1.6	-2.8	N	60	-0.3	-0.5	R
7	-1.4	-2.1	O	61	0.2	-0.5	R
8	-1.4	-2.4	N	62	0.5	-0.2	R
9	-0.9	-2.7	N	63	0.6	-0.7	R
10	-0.7	-2.2	R	64	0.7	-0.4	R
11	-0.5	-2.6	N	65	1.2	-0.3	O
12	-0.2	-2.3	R	66	1.5	-0.5	O
13	0.1	-2.6	N	67	1.7	-0.8	O
14	0.6	-2.2	O	68	1.8	-0.3	O
15	0.6	-2.8	O	69	2.1	-0.3	N
16	0.9	-2.6	R	70	2.2	-0.7	R
17	1.2	-2.3	O	71	2.8	-0.2	N
18	1.4	-2.9	N	72	2.9	-0.6	N
19	1.7	-2.1	O	73	-2.9	0.2	N
20	1.7	-2.5	N	74	-2.6	0.8	O
21	2.2	-2.6	N	75	-2.5	0.2	R
22	2.4	-2.2	N	76	-2.2	0.4	N
23	2.5	-2.8	N	77	-1.7	0.7	O
24	2.7	-2.4	N	78	-1.6	0.2	O
25	-2.9	-1.1	N	79	-1.3	0.5	R
26	-2.8	-1.9	N	80	-1.2	0.2	O
27	-2.6	-1.5	N	81	-0.9	0.8	O
28	-2.1	-1.8	O	82	-0.8	0.4	R
29	-1.8	-1.2	R	83	-0.6	0.6	R
30	-1.7	-1.6	O	84	-0.2	0.6	N
31	-1.4	-1.2	O	85	0.2	0.2	R
32	-1.1	-1.7	O	86	0.2	0.7	R
33	-0.8	-1.5	O	87	0.6	0.8	N
34	-0.6	-1.2	R	88	0.8	0.2	R
35	-0.4	-1.6	O	89	1.2	0.7	O
36	-0.2	-1.1	N	90	1.5	0.4	O
37	0.3	-1.2	O	91	1.8	0.1	O
38	0.3	-1.8	O	92	1.8	0.8	R
39	0.7	-1.2	R	93	2.1	0.3	R
40	0.8	-1.7	O	94	2.3	0.6	N
41	1.2	-1.2	R	95	2.5	0.2	N
42	1.3	-1.8	O	96	2.7	0.8	N
43	1.5	-1.5	O	97	-2.9	1.9	N
44	1.8	-1.7	R	98	-2.7	1.3	N
45	2.3	-1.4	N	99	-2.5	1.7	N
46	2.3	-1.9	N	100	-2.2	1.4	O
47	2.5	-1.2	O	101	-1.9	1.9	R
48	2.9	-1.7	N	102	-1.7	1.5	O
49	-2.8	-0.8	N	103	-1.4	1.8	O
50	-2.8	-0.4	N	104	-1.2	1.1	R
51	-2.3	-0.9	O	105	-0.7	1.4	O
52	-2.2	-0.2	R	106	-0.4	1.2	R
53	-1.7	-0.7	O	107	-0.3	1.6	O
54	-1.6	-0.2	O	108	0.0	1.5	R

Table 4 - continued

Respondent	U	V	Shopping at hypermarket	Respondent	U	V	Shopping at hypermarket
109	0.3	1.3	O	127	-1.5	2.9	N
110	0.4	1.8	O	128	-1.1	2.5	N
111	0.7	1.2	R	129	-0.8	2.9	N
112	0.9	1.6	R	130	-0.6	2.0	O
113	1.3	1.7	O	131	-0.4	2.7	N
114	1.6	1.9	O	132	-0.3	2.5	R
115	1.6	1.6	O	133	0.4	2.3	O
116	1.6	1.2	O	134	0.5	2.9	N
117	2.2	1.7	N	135	0.8	2.1	O
118	2.4	1.3	N	136	0.9	2.6	N
119	2.8	1.9	N	137	1.3	2.8	N
120	2.9	1.5	N	138	1.4	2.6	O
121	-2.8	2.2	N	139	1.5	2.2	N
122	-2.8	2.8	N	140	1.8	2.7	R
123	-2.6	2.5	N	141	2.2	2.9	N
124	-2.2	2.7	N	142	2.4	2.4	N
125	-1.8	2.5	N	143	2.8	2.3	N
126	-1.6	2.1	N	144	2.8	2.9	N

Table 5.

Order of surfaces	Maximized log likelihood	Difference between the maximized log likelihoods of surface orders					Goodness of fit	
			Expected under H_0			Decision		
		Observed	1%	5%	D.F.	on H_0	ρ^2	R^2_p
0	−154.634							
		0.213	6.638	4.744	4	Fail to reject		
1	−154 421						0.0014	0.0033
		47.370	8.406	6.296	6	Reject at 1% level		
2	−107.051						0.3077	0.5475
		3.019	10.045	7.754	8	Fail to reject		
3	−104.032						0.3272	0.5715

the first order surfaces provide an insignificant improvement over the zero order surfaces. The second order surfaces, however, capture a significant amount of the underlying regional structure and it is clearly worthwhile making an exception to the test procedure explained in II (iv) hypothesis b* and extending the models to this higher order. The increased complexity of the probability surface models associated with moving from second order surfaces to third order surfaces is associated with only a minor improvement in the maximized log likelihood, and thus, once again, there is no significant advantage to be gained in fitting third order surfaces. The second order surfaces are therefore the appropriate surfaces to fit and they represent expressions of the underlying regional structure. Figures 10 (a), (b) and (c) (which are Figures 2, 3 and 4 repeated for convenience but without the hand drawn contours) show respectively the second order probability surfaces of regularly shopping at the hypermarket, occasionally shopping at the hypermarket and never shopping at the hypermarket. The reader should note, once again, how the probability surface of regularly shopping at the hypermarket is a 'dome' centred on the hypermarket, how the probability surface of occasionally shopping at the hypermarket has a 'ring doughnut' type of structure with the highest predicted probabilities occurring in a ring or horseshoe some distance from the hypermarket, and how the probability surface of never shopping at the hypermarket is a flat-bottomed 'basin' centred on the hypermarket.

In addition to the inferential tests presented in Table 5, residuals from the probability surface maps should be mapped and an attempt made to assess the amount of systematic spatial variation in these maps. As in the previous example, the program presented in Appendix 1(b) automatically prints both raw residuals (see equation 43) and standardised residuals (see equation 44) and the construction of these maps will be left as an exercise for the reader.

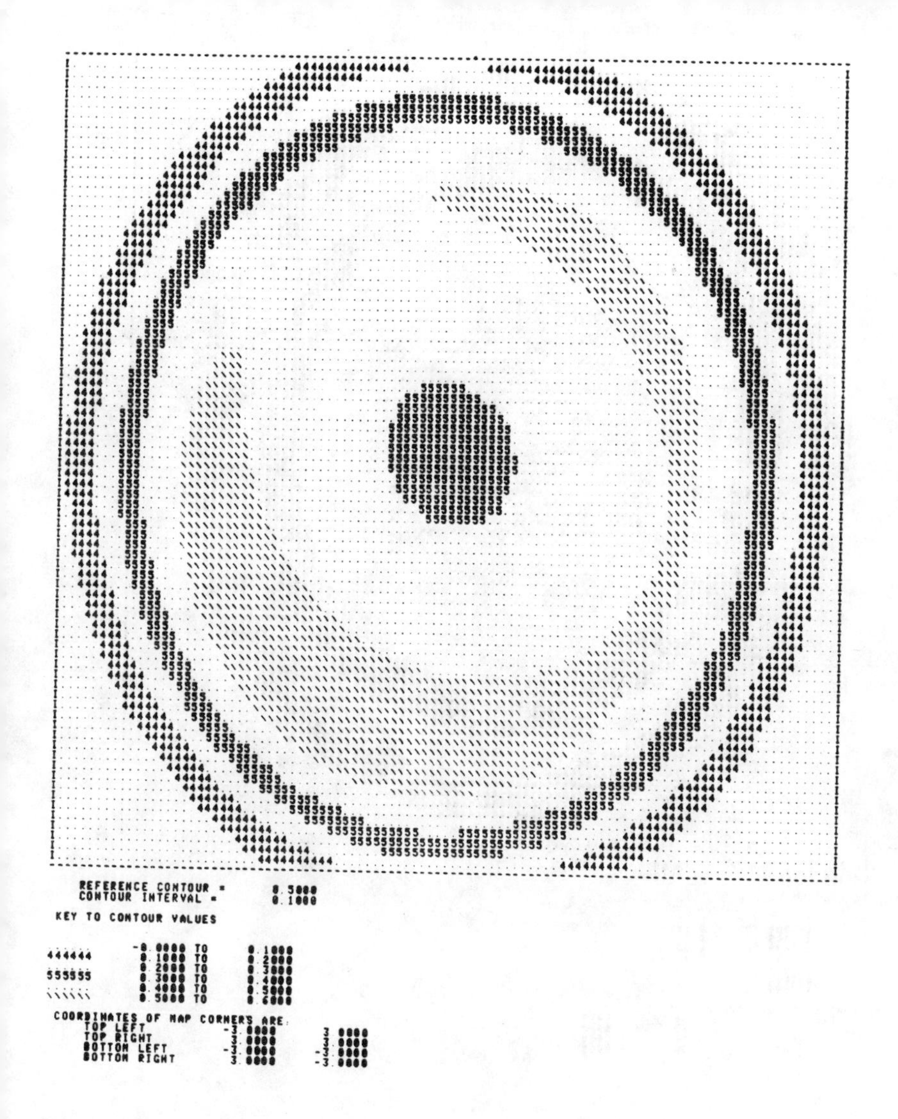

Fig. 10(a) 2nd order probability surface of regularly shopping at the hypermarket

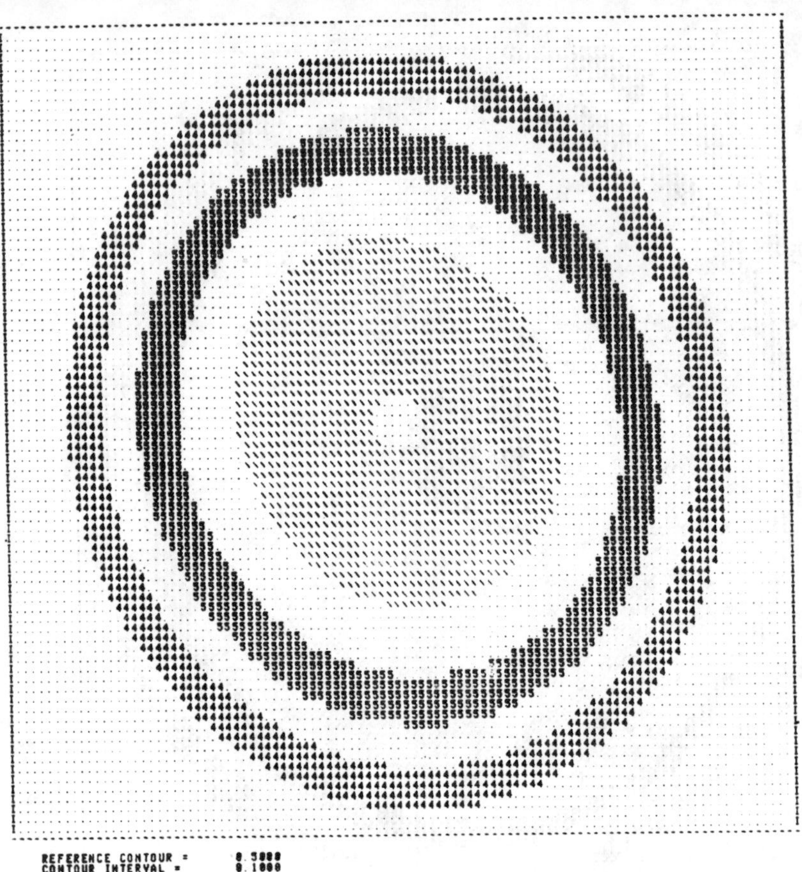

```
REFERENCE CONTOUR =      0.3000
CONTOUR INTERVAL =       0.1000

KEY TO CONTOUR VALUES

......     -0.0000 TO    0.1000
444444      0.1000 TO    0.2000
......      0.2000 TO    0.3000
555555      0.3000 TO    0.4000
......      0.4000 TO    0.5000
\\\\\\      0.5000 TO    0.6000
......      0.6000 TO    0.7000

COORDINATES OF MAP CORNERS ARE:
    TOP LEFT        -3.0000     3.0000
    TOP RIGHT        3.0000     3.0000
    BOTTOM LEFT     -3.0000    -3.0000
    BOTTOM RIGHT     3.0000    -3.0000
```

Fig. 10(b) 2nd order probability surface of occasionally shopping at the hypermarket

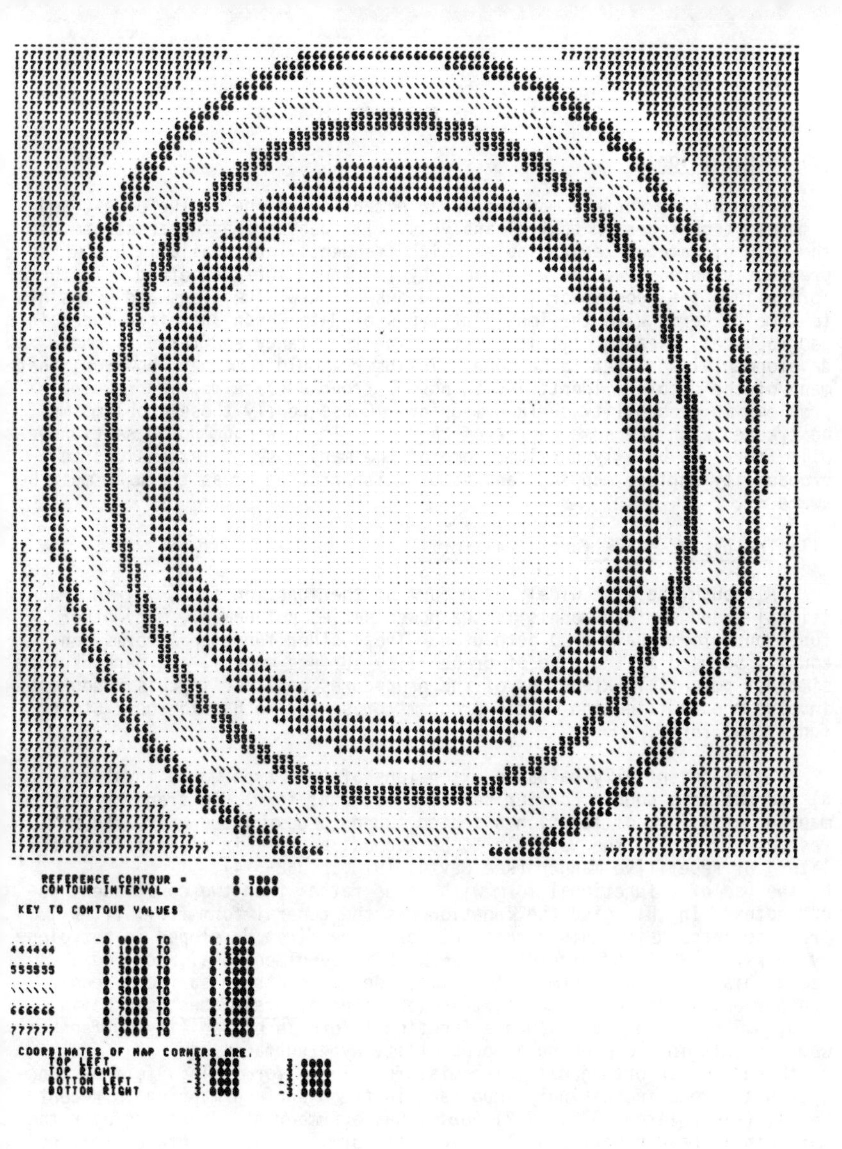

Fig. 10(c) 2nd order probability surface of never shopping at the hypermarket

IV EXTENSIONS AND PROBLEM AREAS

(i) Introduction

Probability surface mapping is an extension of one of the oldest and simplest techniques in spatial analysis. It enables the researcher to use the type of categorized variables which frequently are available but which previously were viewed as unmappable by the trend surface method. It therefore allows the geographer to pose and answer 'new' questions and helps him to come to terms with the imperfect types of data which are often faced in geographical research. It is a method in its infancy and there is much developmental work yet to be done. The user should remember, however, that many of the problems identified in normal trend surface mapping remain problems in probability surface mapping. Consequently these problems must be treated with the same degree of caution. In this concluding section we will therefore briefly consider some of the many possible extensions to probability surface mapping, and some of the problem areas the user must be aware of.

(ii) The use of other functional forms

Although the most widely used form of the function $f(U_i, V_i)$ in traditional trend surface models is the power series polynomial, many other functional forms have also been used. These other functional forms can equally be used in the case of probability surface models, and perhaps the simplest potential extensions of the procedures outlined in this monograph involve the modification of the programs presented to handle these other functional forms.

Three potentially valuable extensions of this type would involve:
a) the use of a double Fourier series functional form. In trend surface mapping this form of function has been found to be of value where the response variable, the variable to be mapped, behaves in a spatially oscillatory or repetitive manner (see Davis, 1973, p.358-374).
b) the use of a functional form with three rather than two constituent co-ordinates. In this case the function has the general form $f(U_i, V_i, W_i)$. Trend surfaces using this functional form were first developed in petrology by Peikert (1962) and others for use with three-dimensional rock bodies. Using this functional form contour lines on normal trend surfaces become contour envelopes, and the effect is to define <u>hypersurfaces</u> (see Davis, 1973, p.355). The use of such a functional form in probability surfaces would result in the mapping of probability hypersurfaces.
c) the use of an orthogonal polynomial functional form. This form of function has become increasingly important in trend surface mapping in recent years, (see Whitten 1970, 1972) for it has a number of advantages over the more widely used nonorthogonal polynomial form. These advantages include the added numerical accuracy it is possible to achieve in the computation of the parameter estimates (see Mather, 1976, p.139-142) and the provision of information concerning the importance of each orthogonal parameter.

(iii) Programs for fitting different surfaces to different responses

Probability surface models are essentially sets of linked models; two linked models in the two response category case producing two probability surface maps, three linked models in the three response category case producing three probability surface maps and so on. The predicted probabilities at any locality i in each set of maps sum to one. To produce the J probability surface maps in a J response category case, the programs presented in Appendix 1(a), 1(b) and 1(c) fit the same order of surface to each response. A logical extension of these programs would be to allow a different order of surface to be fitted to each of the J responses, whilst retaining the property that the predicted probabilities of the J responses at any locality must sum to one. This extension is simply a computer programming problem, it involves no alterations to the structure of the probability surface models outlined above.

(iv) Ordered categorizations

Throughout the monograph no account has been taken of any ordering which might exist amongst the response categories. A potential extension to the probability surface models outlined is to attempt to take account in the specification of the models and in the estimation procedures of any ordering which might exist amongst the response categories. Cox (1970, p.104) presents a nonlinear model which takes account of the ordering of the responses in a particular three category situation.

(v) Inference and estimation considerations

In II (iv) and (v) we discussed the need for a test statistic to evaluate the amount of systematic spatial variation in the maps of residuals from probability surfaces of different orders, and the need to resolve what constitutes an appropriate goodness-of-fit statistic. Although these are perhaps the most urgent inferential matters to be resolved, consideration must also be given to the question of how appropriate is the maximum likelihood estimation method and inferential tests (see equation 28) based upon it, in the case of small samples. Some limited Monte Carlo experiments described by Domencich and McFadden (1975, p.112-117) suggest that the maximum likelihood estimators perform reasonably well when compared with the alternative weighted least squares estimators in small samples, but further small sample Monte Carlo experiments would be useful.

(vi) Graphical improvements

Although the line printer maps produced by the programs given in Appendix 1(a), 1(b) and 1(c) provide a quick and reasonably accurate method of displaying probability surfaces, there is clearly considerable potential for improving the quality of the maps produced, either by using more elaborate line printer techniques, or drum or flat bed plotter methods (see Peuker, 1972; Davis and McCullagh, 1975; Rhind, 1977). The additional use of three-dimensional perspective views of the probability surfaces would also help to improve the graphical output from the programs presented.

(vii) Spatial distribution of the data points

In most text book discussions of traditional trend surface mapping the user is warned that the spatial arrangement of the data points can have profound effects on the shape of the computed trend surfaces (Davis, 1973, p.349-352; Mather, 1976, p.120-130; Unwin, 1975a, p.31-32). Given that there are at least as many, preferably many more, data points than there are parameters to be estimated in the trend surface model, the user is warned that the data points should have an even spatial distribution, and that the map area should be approximately square.

If these two conditions are not satisfied, the shape of the trend surface fitted can be greatly distorted. For example, if there are few data points at the edges of the map area, the fitted surfaces lack constraints on their form in these areas. Whatever slope exists in the region of the map in which there is data point control, is extrapolated without limits along the map edges. This creates what are termed 'edge effects', and these effects can be very serious in the case of higher order surfaces. To guard against these effects it is wise to form a 'buffer region' (Davis, 1973, p.350) of control points around the map, so that edge effects can be concentrated into this buffer region, leaving the area of interest with adequate data point control. In addition to having sufficient data point control at the edges of the map, the user must also take care that the data point distribution is in no way peculiar, for example, highly clustered or restricted to a narrow strip across the map area (Doveton and Parsley, 1970). He should also ensure that the map area is not markedly rectangular in shape. If it is there will be a pronounced tendency for contours on higher order surfaces to become elongated parallel to the long axis of the map area.

These problems associated with the spatial distribution of the data points are potentially as great a threat to the user of probability surface mapping as they are to the user of trend surface mapping. The user of probability surface mapping should therefore attempt to ensure an even spatial distribution of data points and an approximately square map area. If he is in the position of being able to construct a sample design for the collection of data to be used in probability surface mapping, he should consider using the spatially stratified random sampling method which underlies Figures 5 and 9.

(viii) The problem of multicollinearity

In trend surface mapping, accurate calculation of the parameter estimates of traditional polynomial trend surface models is difficult, for the matrix of sums of squares and cross-products of the explanatory variables (the U_i, V_i terms) is almost certain to be nearly singular for surfaces of order three or more. This results from the fact that the explanatory variables will be approximately linearly related, or in other words because of the presence of a high degree of multicollinearity.

In the presence of such a high degree of multicollinearity, the ordinary least squares estimation methods employed by many of the standard trend surface computer programs produce inaccurate results, particularly when the problem is compounded by a poor data point distribution, a poorly selected origin for the co-ordinate system, and a computer with a short word length (see Unwin, 1975b; Mather, 1977). Such problems can be minimized by using the orthogonal polynomials mentioned in IV (ii) above, by the choice of a sensible origin for the co-ordinate system and/or scaling of the explanatory variables, by choice of a sampling design which produces a better data point distribution and the use of a computer with as long a word as possible.

In the case of probability surface mapping we use a maximum likelihood method of parameter estimation, however, a high degree of multicollinearity still causes problems. Cox (1970, p.90) for example, discussing maximum likelihood estimation of nonlinear logit models, states that 'there is likely to be difficulty in finding (the parameter estimates) if the columns of the (matrix of explanatory variables) are nearly linearily dependent. Therefore it may be good to have a preliminary calculation of the formal 'correlation' matrix of the regressor (i.e. explanatory) variables, followed if necessary by a linear transformation of the regressor variables to ones more nearly orthogonal in the usual least squares sense'. In this monograph no such transformation has been conducted in the examples given, for a sensible origin for the co-ordinate system was chosen, the sampling design employed produced a good data point distribution, and the programs presented have been restricted to fitting surfaces no higher than fourth order. No difficulties were encountered in finding the parameter estimates. In less favourable circumstances, however, or if the user felt that it was absolutely essential to fit probability surfaces of higher order, the use of orthogonal polynomials, and the extension of the computer programs presented here to handle them, appears to be a possible way forward. Even in the absence of such a need, however, it would be interesting to compare the results presented in II (vi) and III (iv) based on the more traditional nonorthogonal polynomials, with those achieved using orthogonal polynomials.

(ix) Conclusion

Probability surface mapping extends one of the oldest, simplest and most widely known techniques in spatial analysis, and it offers the geographer a method capable of handling the type of categorized data which are often faced in geographical research. The aim of this monograph has been to make the method available to all those who require it, and it is hoped that this will stimulate further empirical application, theoretical development, and programming refinement.

BIBLIOGRAPHY

A. Regression and trend surface analysis

Bassett, K. (1972), Numerical methods for map analysis in: *Progress in Geography,* 4, eds. C. Board et al. (London, Arnold).

Chorley, R.J. and Haggett, P. (1965), Trend surface mapping in geographical research. *Transactions Institute of British Geographers* 37, 47-67.

Cliff, A.D. and Ord, J.K. (1969), The problem of spatial autocorrelation in: London papers in regional science, 1, *Studies in regional science,* ed. A.J. Scott, (London, Pion).

Cliff, A.D. and Ord, J.K. (1972), Testing for spatial autocorrelation among regression residuals. *Geographical Analysis,* 4, 267-284.

Cliff, A.D. and Ord, J.K. (1973), *Spatial autocorrelation.* (London, Pion).

Davis, J.C. (1973), *Statistics and data analysis in geology.* (New York, Wiley).

Doveton, J.H. and Parsley, A.J. (1970), Experimental evaluation of trend surface distortions induced by inadequate data point distributions. *Transactions, Institute of Mining and Metallurgy.* Section B, B197-208.

Ferguson, R.I. (1977), *Linear regression in geography.* Concepts and techniques in modern geography, 15. (Norwich, Geo Abstracts Ltd).

Huang, D.S. (1970), *Regression and econometric methods.* (New York, Wiley).

Kmenta, J. (1971), *Elements of econometrics.* (New York, Macmillan).

Mather, P.M. (1976), *Computational methods of multivariate analysis in physical geography.* (London, Wiley).

Mather, P.M. (1977), Clustered data-point distributions in trend surface analysis. *Geographical Analysis,* 9, 84-93.

Peikert, E.W. (1962), Three dimensional specific gravity variations in the Glen Alpine Stock, Sierra Nevada, California. *Geological Society of America Bulletin,* 73, 1437-42.

Unwin, D.J. (1975a), *An introduction to trend surface analysis.* Concepts and techniques in modern geography, 5. (Norwich, Geo Abstracts Ltd).

Unwin, D.J. (1975b), Numerical errors in a familiar technique: a case study of polynomial trend surface analysis. *Geographical Analysis,* 7, 197-203.

Whitten, E.H.T. (1970), Orthogonal polynomial trend surfaces for irregularly-spaced data. *Journal of the International Association for Mathematical Geology,* 2, 141-52.

Whitten, E.H.T. (1972), More on 'irregularly-spaced data and orthogonal polynomial trend surfaces'. *Journal of the International Association for Mathematical Geology,* 4, 83.

Whitten, E.H.T. (1975), The practical use of trend-surface analyses in the geological sciences. in: Davis, J.C. and McCullagh, M.J. (eds.). *Display and analysis of spatial data.* (London, Wiley).

Wonnacott, R.J. and Wonnacott, T.H. (1970), *Econometrics.* (New York, Wiley).

B. Statistical analysis of categorized variables

Baxter, N.D. and Cragg, J.G. (1970) Corporate choice among long-term financing instruments. *Review of Economics and Statistics* 52, 225-35.

Berkson, J. (1953), A statistically precise and relatively simple method of estimating the bio-assay with quantal response based on the logistic function. *Journal American Statistical Association,* 48, 565-99.

Cox, D.R. (1970), *The analysis of binary data.* (London, Methuen).

Domencich, T.A. and McFadden, D. (1975), *Urban travel demand. A behavioural analysis.* (Amsterdam, North-Holland).

Grizzle, J.E., Starmer, C.F. and Koch, G.G. (1969), Analysis of categorical data by linear models. *Biometrics,* 25, 489-504.

Mantel, N. (1966), Models for complex contingency tables and polychotomous dosage response curves. *Biometrics,* 22, 83-95.

Mantel, N. and Brown, C. (1973), A logistic re-analysis of Ashford and Sowden's data on respiratory symptoms in British coal miners. *Biometrics,* 29, 649-65.

McFadden, D. (1974), Conditional logit analysis of qualitative choice behaviour in: *Frontiers in econometrics,* (ed) P. Zarembka, (New York, Academic Press).

Pindyck, R.S. and Rubinfeld, D.L. (1976), *Econometric models and economic forecasts.* (New York, McGraw-Hill).

Richards, M.G. and Ben-Akiva, M.E. (1975), *A disaggregate travel demand model.* (Farnborough, Saxon House).

Schmidt, P. and Strauss, R. (1975), The prediction of occupation using multiple logit models. *International Economic Review,* 16, 471-86.

Tardiff, T.J. (1976), A note on goodness-of-fit statistics for probit and logit models. *Transportation,* 5, 377-88.

Theil, H. (1970), On the estimation of relationships involving qualitative variables. *American Journal of Sociology,* 76, 103-54.

Wrigley, N. (1975), Analysing multiple alternative dependent variables. *Geographical Analysis,* 7, 187-95.

Wrigley, N. (1976), *An introduction to the use of logit models in geography.* Concepts and techniques in modern geography, 10. (Norwich, Geo Abstracts Ltd).

Wrigley, N. (1977), Probability surface mapping: a new approach to trend surface mapping. *Transactions Institute of British Geographers,* New Series, 2, 129-40.

C. Other references

Davis, J.C. and McCullagh, M.J. (1975), (eds) *Display and analysis of spatial data.* (London, Wiley).

Edwards, A.W.F. (1972), *Likelihood.* (Cambridge, U.P.).

Peuker, T.K. (1972), Computer cartography. *Association of American Geographers, Commission on College Geography.* Resource Paper No. 17.

Rhind, D. (1977), Computer-aided cartography. *Transactions Institute of British Geographers,* New Series, 2, 71-97.

APPENDIX 1 Computer programs for probability surface mapping

Fortran programs for the two, three and four response category cases are provided in this Appendix. All the programs share certain common subroutines which are listed in (d) below, and all allow the user to fit, in a sequential manner, polynomial probability surfaces from order 1 up to order 4. It should be noted in this context that although users of traditional trend surface mapping have fitted polynomial trend surfaces more complex than 4th order, and although the programs provided in the Appendix could be extended to allow similar fitting of surfaces more complex than 4th order, this facility is not provided because there are marked dangers, both theoretical and numerical, associated with the fitting of such higher order surfaces.

The programs provided in this Appendix must be viewed as first generation probability surface programs, capable of considerable future refinement and extension. They are presented simply to allow the method to break through what might otherwise prove to be an initial computational barrier to its wider implementation. Readers so inclined, are encouraged to develop and refine the programs presented here.

Appendix 1(a) A Fortran program for the two response category case

The program presented in Table 6 has the following stages.

a) Input It starts by reading a card which specifies the highest order of surface the user wishes to fit (IORD); the number of respondents in the sample (NSAMPL), and the number of respondents who have chosen response category one, in terms of the example in II (vi) the number who have claimed to shop at the hypermarket (N1). It then reads another card which specifies the overall dimensions the user wishes the probability surfaces to be drawn to (10.0 inches by 10.0 inches in the example in II (vi) before reduction to the monograph page size); the U_i and V_i values of the corners of the maps; the contour interval required (0.10 in the example), and the so called reference contour, a contour which should be set to a figure in the middle of the expected range of predicted probabilities (set at 0.50 in the example for convenience). Following this it reads a set of cards. These give the geographical co-ordinates of the sample localities (see Table 1). The input order required of these cards demands that all respondents choosing response category one form the first N1 cards. The next set of cards N1+1 to NSAMPL are the respondents choosing response category two. In the example in II (vi) Table 1 must therefore be reorganized so that the first N1 cards are the housewives claiming to shop at the hypermarket, whilst the next set of cards N1+1 to NSAMPL are the housewives claiming not to shop at the hypermarket. This completes the input.
b) Calculations After receiving the input information the program then fits in a sequential manner, probability surfaces of order 1 to IORD. For each order of probability surface the parameter estimates of the probability surface models are estimated using the maximum likelihood method, and these parameter estimates are then used as the basis of a line printer subroutine which draws the probability surface maps.
c) Output The geographical co-ordinates of the sample households in the order in which they were input, see (a) above, are listed first. Then for each order 1 to IORD, the following is provided: the maximized log likeli-

hood, parameter estimates, standard errors, variance-covariance matrix of the parameter estimates, predicted probabilities of giving each response for each of the sample respondents, raw residuals, standardised residuals, and finally the probability surface maps for both the possible responses. Prior to this the maximized log likelihood for the probability surface models of order 0, that is to say for the probability surface models containing only an intercept term, is printed.

The program given in Table 6 has the following size restrictions. The maximum number of respondents in the sample (NSAMPL) allowed is 250. The maximum number who choose response category one (N1) allowed (i.e. the number who have claimed to shop at the hypermarket) is 125. The maximum number who choose response category two (NSAMPL-N1) allowed (i.e. the number who have claimed not to shop at the hypermarket) is 200. These size restrictions are easily changed however and in Appendix 2(a) the changes required in the program when NSAMPL = AT1, N1 = AT2 and (NSAMPL-N1) = AT3 (where AT1, AT2, and AT3 represent any arbitrary numbers) are given, thus allowing the user to modify the size restrictions of the program to suit his own needs.

Before using the program for his own research, the reader is advised to check, using the example data set of Table 1, that he has punched and implemented the program correctly. Minor differences in numerical results from those given in the next section can be expected to occur if the computer used by the reader has a level of precision which differs from that of the University of Bristol machine.

After ensuring that the program has been implemented correctly and reproduces the results reported in II (vi), other possible (but rare) errors which the user may encounter are as follows.

a) Program exits from subroutine VA06AD with message 'Error exit from VA06AD'. The subroutine may finish in this way because the gradients are wrong or because computer rounding errors make it impossible to continue the calculation efficiently. If this message occurs and if the final derivative vector (the G vector printed immediately before the 'Maximized Log Likelihood Value' statement) does not have small components, then it is probably due to incorrect derivatives. The user should note that this error has never been encountered by the author and should check that he has implemented the program correctly.

b) Program exits from subroutine VA06AD with message 'VA06AD has made 2000 calls of CALCFG'. The program has failed to achieve a maximum after 2000 iterations (the arbitrarily set upper limit). The program normally achieves a maximum in less than max(100, 10xn) where n here is the total number of parameters in the set of probability surface models, and therefore this suggests serious problems caused by linear dependence between the explanatory variables. The user should read Section IV (viii). He should also, check that additional comment instruction (b) given in the next paragraph has been obeyed, and try following the instructions given in (c) below to see if this improves the position.

c) Negative square root problem immediately after VA06AD. The user should change the arbitrarily set initial parameter estimates given in lines 33 and 35 of subroutine SURF. It is suggested that line 35 be changed from X(I)=0.01D0 to X(I)=0.02D0, or X(I)=0.03D0.

Additional comments.

a) The probability surface maps are printed on the line printer. The pro-

gram assumes that the line printer prints 8 lines per vertical inch. The user should check what number of lines per vertical inch are printed by his local line printer. In many computer centres 6 lines per vertical inch may be the norm. If this is the case, the user should change line 24 in the main part of the program from NL=INT(VERTx8.0+0.5) to NL=INT(VERTx6.0+0.5).

b) Subroutine VA06AD works most efficiently when the observed values of the explanatory variables differ in scale by a factor of no more than 100. In the polynomial probability surface case the kth order models have a range of values from U_i to U_i^k and V_i to V_i^k. To ensure that this range of values approximates to the range in which subroutine VA06AD works most efficiently, it is essential that the origin (the zero point) of the co-ordinate system coincides with the centre of the study area (see Figure 5). Also the units in which U_i and V_i are expressed should follow the pattern of Figure 5. That is to say, ideally there should be only one number to the left of the decimal point.

c) The DATA statement in subroutine PMAP sets the characters to be used in the line printer maps (see Figures 6, 7 and 8, III (vi)). The user should note that the small dot symbol is used for every other probability division to aid the visual appearance of the maps. He should also note that the eleventh probability division is denoted in the DATA statement and in Figures 6, 7 and 8, by the symbol\ . Users unable to punch this symbol at their local computer centre should replace it in the DATA statement with a $ symbol.

```
****** PROBABILITY SURFACE PROGRAMME FOR A 2 CATEGORY SITUATION
WRITTEN BY NEIL WRIGLEY, DEPT  OF GEOGRAPHY, UNIVERSITY OF BRISTOL

COMMON/ONE/A(250,15),A2(125,15),A1(200,15),NSAMPL,N1,N1A,K,N4B

***** READ IN IORD=HIGHEST SURFACE ORDER TO BE FITTED, 1,2,3, OR 4
NSAMPL=NUMBER OF SAMPLE LOCALITIES, N1=NUMBER OF LOCALITIES WHERE
RESPONSE CATEGORY ONE IS RECORDED *********************************
**** DATA CARD: COL 1=IORD, COL 2=BLANK, COL 3-5=NSAMPL, COL 6=
BLANK, COL 7-9=N1 ***********************************************
READ(5,1) IORD,NSAMPL,N1
FORMAT(I1,1X,I3,1X,I3)

**** READ IN VERT=LENGTH OF PROBABILITY SURFACE MAP REQUIRED IN
INCHES, WID=WIDTH IN INCHES, REFC=VALUE OF REFERENCE CONTOUR, CINT
=CONTOUR INTERVAL, UMAX,UMIN=MAXIMUM AND MINIMUM ALONG HORIZONTAL
AXIS, VMAX,VMIN=MAXIMUM AND MINIMUM ALONG VERTICAL AXIS *********
*** DATA CARD: COLS 1-7=VERT, COLS 8-14=WID, COLS 15-19=REFC,
COLS 20-24=CINT, COLS 25-31=UMAX, COLS 32-38=UMIN, COLS 39-45=VMAX
COLS 46-52=VMIN *************************************************
READ(5,10) VERT,WID,REFC,CINT,UMAX,UMIN,VMAX,VMIN
FORMAT(2F7.4,2F5.3,4F7.4)
NL=INT(VERT*8.0+0.5)
NC=INT(WID*10.0+0.5)
UAX=(UMAX-UMIN)/FLOAT(NC-1)
VAX=(VMAX-VMIN)/FLOAT(NL-1)
UAX=UAX+0.000001*UAX
VAX=VAX+0.000001*VAX

**** READ IN GEOGRAPHICAL CO-ORDINATES OF EACH SAMPLE LOCALITY,
U(I)=HORIZONTAL CO-ORDINATE, V(I)=VERTICAL CO-ORDINATE, IN FORMAT
(F8.4,1X,F8.4) (N.B. IN THE PROGRAMME THESE CO-ORDINATES ARE
STORED IN MATRIX A) *** ORDER OF CARDS-ALL SAMPLE LOCALITIES WHERE
RESPONSE CATEGORY ONE IS RECORDED, FOLLOWED BY ALL LOCALITIES
WHERE REMAINING RESPONSE CATEGORY IS RECORDED *** CREATE CONSTANT
TERM *** PRINT SAMPLE LOCALITIES *********************************
WRITE(6,11)
FORMAT('0       U            V')
DO 3 I=1,NSAMPL
A(I,1)=1.0
READ(5,4) A(I,2),A(I,3)
WRITE(6,44) (A(I,J),J=2,3)
FORMAT(F8.4,1X,F8.4)
FORMAT(' ',2F9.4)

***** START OF MAIN LOOP ****************************************
ID=0
DO 7 I=1,IORD
ID=ID+1
CALL SURF(ID,NL,NC,UAX,VAX,UMAX,UMIN,VMAX,VMIN,REFC,CINT)
CONTINUE
STOP
END




SUBROUTINE SURF(IORD,NL,NC,UAX,VAX,UMAX,UMIN,VMAX,VMIN,REFC,CINT)
COMMON/ONE/A(250,15),A2(125,15),A1(200,15),NSAMPL,N1,N1A,K,N4B
DOUBLE PRECISION BZ1,DBLE
DOUBLE PRECISION X,F,G,STEP,ACC,W,FMINUS,STER,TSTAT
DIMENSION C10(250),C13(250)
DIMENSION PROB1(250),PROB4(250),QZA(250,1),XZ1(15,1)
DIMENSION X(15),G(15),TSTAT(15),STER(15),WORK(15),PESTS(15),W(541)

*** THIS IS THE MAIN PART OF THE PROGRAMME. IT CALCULATES THE
PARAMETER ESTIMATES OF THE PROBABILITY SURFACE MODEL USING MAXIMUM
LIKELIHOOD ESTIMATION. IT USES POWELL'S HYBRID STEEPEST DESCENT
AND GENERALISED NEWTON METHOD ***********************************

**** FOLLOWING ARBITRARILY SETS THE ITERATION INSTRUCTIONS *****
MAXFUN=2000
IPRINT=20
```

```
      BZ=FLOAT(IORD)
      BZ1=DBLE(BZ*0.03)
      STEP=BZ1+0.03D0
      IF(IORD.EQ.1) ACC=0.08D0
      IF(IORD.EQ.2) ACC=0.16D0
      IF(IORD.EQ.3) ACC=0.27D0
      IF(IORD.EQ.4) ACC=0.40D0
C
C     ******* CALCULATE NUMBER OF PARAMETERS *************************
      K=((IORD+1)*(IORD+2))/2
      N=K
      N1A=N1+1
      N40=NSAMPL-N1
      J92=0
C
C     ****** GENERATE ARBITRARY INITIAL PARAMETER ESTIMATES ********
      X(1)=1.0D0
      DO 2 I=2,N
2     X(I)=0.1D0
C
C     ******* GENERATE POLYNOMIAL TERMS ******************************
      IF(IORD.EQ.1) GO TO 12
      KD=IORD-1
      KDA=((KD+1)*(KD+2))/2
      KDB=KDA+1
      KE=IORD-2
      KEA=((KE+1)*(KE+2))/2
      KEB=KEA+1
      DO 15 J=KEB,KDA
      DO 14 I=1,NSAMPL
14    A(I,KDB)=A(I,J)*A(I,2)
      KDB=KDB+1
15    CONTINUE
      DO 16 I=1,NSAMPL
      A(I,KDB)=A(I,KDA)*A(I,3)
16    CONTINUE
12    CONTINUE
C
C     ***** CALCULATE PARAMETER ESTIMATES ****************************
      DO 5 I=1,N1
      DO 5 J=1,K
5     A2(I,J)=A(I,J)
      DO 8 I=N1A,NSAMPL
      J92=J92+1
      DO 8 J=1,K
8     A1(J92,J)=A(I,J)
      IF(IORD.NE.1) GO TO 8690
      W18=FLOAT(NSAMPL)
      W19=FLOAT(N1)
      W20=(W18-W19)/W18
      W21=W18-W19
      W22=W21*ALOG(W20)
      W23=W19/W18
      W25=W19*ALOG(W23)
      W26=W25+W22
      WRITE(6,8691)
 8691 FORMAT(/////,1X,'FOLLOWING IS THE MAXIMIZED LOG LIKELIHOOD VAL
     1OR INTERCEPT ONLY MODEL')
      WRITE(6,369) W26
 8690 WRITE(6,6200) IORD
 6200 FORMAT(////////,1X,'PROBABILITY SURFACE MODEL OF ORDER ',I1,'
     1FITTED AND SURFACES DRAWN')
      CALL VA06AD(N,X,F,G,STEP,ACC,MAXFUN,IPRINT,W)
      FMINUS=-1.0D0*F
      WRITE(6,369) FMINUS
369   FORMAT(//,1X,'MAXIMIZED LOG LIKELIHOOD VALUE = ',F12.5)
      NW12=0.5*N*(N+1)
      NW13=NW12+1
      NLESS=N-1
      N70=NW13
      N200=NW13+NLESS
      DO 6001 K2004=1,N
      N71=NLESS-K2004
      STER(K2004)=DSQRT(W(N70))
      N70=N200+1
 6001 N200=N70+N71
      WRITE(6,2008)
 2008 FORMAT(//,1X,'PARAMETER ESTIMATES  STANDARD ERRORS  PARAMETER
     1STANDARD ERROR')
      DO 2007 I=1,N
      TSTAT(I)=X(I)/STER(I)
 2007 WRITE(6,2009) X(I),STER(I),TSTAT(I)
```

```
2009 FORMAT(1X,F15.7,F18.7,F22.7)
     WRITE(6,6009)
6009 FORMAT(//,1X,'VARIANCE-COVARIANCE MATRIX OF PARAMETER ESTIMATES')
8001 FORMAT('-',3X,'COLUMN',I6,9I11)
8003 FORMAT(//,1X,'ROW'/)
8002 FORMAT(' ','ROW ',I3,3X,10F11.5)
     IBEGIN=NW13-1-N
     DO 7002 I=1,N,10
     ILESS1=I-1
     IEND=I+9
     IF(IEND.GT.N) IEND=N
     WRITE(6,8001) (ICOL,ICOL=I,IEND)
     WRITE(6,8003)
     DO 7002 J=I,N
     MAX=J
     IF(MAX.GT.IEND) MAX=IEND
     IBASE=IBEGIN+J
     DO 7001 K17=I,MAX
     INDEX=IBASE+K17*N-(K17*K17-K17)/2
     L=K17-ILESS1
7001 PROB1(L)=SNGL(W(INDEX))
     MAX=MAX-ILESS1
7002 WRITE(6,8002) J,(PROB1(L),L=1,MAX)
     DO 630 I=1,K
630  XZ1(I,1)=SNGL(X(I))
     ******  COMPUTE ESTIMATED PROBABILITIES ************************
     CALL MC01AS(A,XZ1,QZA,NSAMPL,K,1,250,15,250)
     DO 633 I=1,NSAMPL
     QZA(I,1)=EXP(QZA(I,1))
     ZZ=1.0+QZA(I,1)
     PROB1(I)=QZA(I,1)/ZZ
633  PROB4(I)=1.0/ZZ
     DO 8910 I=1,NSAMPL
     IF(PROB1(I).EQ.1.0) PROB1(I)=0.9999999
     IF(PROB1(I).EQ.0.0) PROB1(I)=0.0000001
     IF(PROB4(I).EQ.1.0) PROB4(I)=0.9999999
     IF(PROB4(I).EQ.0.0) PROB4(I)=0.0000001
     C10(I)=SQRT(PROB1(I)*(1.0-PROB1(I)))
     C13(I)=SQRT(PROB4(I)*(1.0-PROB4(I)))
 910 CONTINUE
     WRITE(6,8901)
 901 FORMAT(///,1X,'FOR EACH SAMPLE LOCALITY'/'PREDICTED PROBABILITES
    1RAW RESIDUALS                STANDARDISED RESIDUALS')
     WRITE(6,8930)
 930 FORMAT(/,1X,'RESP CAT 1  RESP CAT 2  RESP CAT 1  RESP CAT 2  RESP
    1CAT 1  RESP CAT 2')
     DO 8920 I=1,N1
     RAWR1=(1.0-PROB1(I))
     RAWR4=-PROB4(I)
     RES1=RAWR1/C10(I)
     RES4=RAWR4/C13(I)
 920 WRITE(6,8902) PROB1(I),PROB4(I),RAWR1,RAWR4,RES1,RES4
     DO 8921 I=N1A,NSAMPL
     RAWR1=-PROB1(I)
     RAWR4=(1.0-PROB4(I))
     RES1=RAWR1/C10(I)
     RES4=RAWR4/C13(I)
 921 WRITE(6,8902) PROB1(I),PROB4(I),RAWR1,RAWR4,RES1,RES4
 902 FORMAT(' ',F8.5,4X,F8.5,4X,4(F9.5,3X))
     DO 243 I=1,K
 43  PESTS(I)=XZ1(I,1)
     CALL PMAP(NL,NC,PESTS,WORK,REFC,CINT,UAX,VAX,N,IORD,VMAX,UMIN,UMAX
    1,VMIN)
     RETURN
     END


     SUBROUTINE CALCFG(N,X,F,G)
     DOUBLE PRECISION X,G,U12,U13,F,DBLE
     COMMON/ONE/A(250,15),A2(125,15),A1(200,15),NSAMPL,N1,N1A,K,N48
     DIMENSION X(15),G(15)
     DIMENSION XN1(15,1),Q1(125,1),U6(250,1),U9(250,1),ROW1(1,250)
     DIMENSION ROW2(1,125),U11(250,15),U12(1,15),U13(1,15)
     DO 66 I=1,K
     XN1(I,1)=SNGL(X(I))
     U1=0.0
```

```
      Z1=0.0
      CALL MC01AS(A2,XN1,Q1,N1,K,1,125,15,125)
      DO 169 I=1,N1
  169 U1=U1+Q1(I,1)
      Z=U1
      CALL MC01AS(A,XN1,U6,NSAMPL,K,1,250,15,250)
      DO 87 I=1,NSAMPL
   87 U6(I,1)=EXP(U6(I,1))
      DO 90 I=1,NSAMPL
      U9(I,1)=U6(I,1)
   90 U9(I,1)=1.0+U9(I,1)
      DO 91 I=1,NSAMPL
   91 Z1=Z1+ALOG(U9(I,1))
      FA=-1.0*(Z-Z1)
      F=DBLE(FA)
C     ***** -F IS THE LOG LIKELIHOOD VALUE ****************************
C
C     *** CALCULATE 1ST DERIVATIVES OF LOG LIKELIHOOD FUNCTION *******
      DO 44 I=1,NSAMPL
   44 ROW1(1,I)=1.0
      DO 45 I=1,N1
   45 ROW2(1,I)=1.0
      DO 94 I=1,NSAMPL
   94 U6(I,1)=U6(I,1)/U9(I,1)
      DO 92 J=1,K
      DO 92 I=1,NSAMPL
   92 U11(I,J)=A(I,J)*U6(I,1)
      CALL MATNW(ROW1,U11,U12,1,NSAMPL,K,1,250,1)
      CALL MATNW(ROW2,A2,U13,1,N1,K,1,125,1)
      DO 93 J=1,K
   93 U13(1,J)=U13(1,J)-U12(1,J)
      DO 101 I=1,K
  101 G(I)=U13(1,I)
      DO 104 I=1,N
  104 G(I)=-G(I)
C     **** G IS THE VECTOR OF 1ST DERIVATIVES ************************
      RETURN
      END




      SUBROUTINE PMAP(NL,NC,PESTS,WORK,REFC,CINT,UAX,VAX,N,IORD,VMAX,
     1N,UMAX,VMIN)
      DIMENSION PESTS(N),WORK(N),SLINE(120),SYMB(23)
      DATA SYMB/1H1,1H.,1H2,1H.,1H3,1H.,1H4,1H.,1H5,1H.,1H\,1H.,1H6,1
     11H7,1H.,1H8,1H.,1H9,1H.,1H0,1HI,1H-/
    1 FORMAT('1       PROBABILITY SURFACE OF ORDER ',I8/' PART',I4//)
    2 FORMAT(' ',120A1)
      DO 66 JZ=1,2
      KMAX=0
      KMIN=20
      NO2=110
      NSTR=NC/110
      LEFT=NC-110*NSTR
      IF(LEFT.GT.0) NSTR=NSTR+1
      DO 96 KOUNT=1,NSTR
      IF(KOUNT.EQ.NSTR.AND.LEFT.GT.0) NO2=LEFT
      WRITE(6,1) IORD,KOUNT
      NO22=NO2+2
      WRITE(6,2) (SYMB(23),I=1,NO22)
      V=VMAX+VAX
      DO 12 L=1,NL
      V=V-VAX
      U=UMIN-UAX+(KOUNT-1)*110*UAX
      DO 6 I=1,NO2
      U=U+UAX
      PR=PROB(PESTS,WORK,N,IORD,U,V,JZ)
      KCC=11+INT((PR-REFC)/CINT)
      IF(PR.LT.REFC) KCC=KCC-1
      IF(KCC.LT.1) KCC=1
      IF(KCC.GT.20) KCC=20
      IF(KCC.GT.KMAX) KMAX=KCC
      IF(KCC.LT.KMIN) KMIN=KCC
    6 SLINE(I)=SYMB(KCC)
   12 WRITE(6,2) SYMB(22),(SLINE(I),I=1,NO2),SYMB(22)
      WRITE(6,2) (SYMB(23),I=1,NO22)
   96 CONTINUE
```

```
      WRITE(6,4) REFC,CINT
4     FORMAT('0    REFERENCE CONTOUR = ',F11.4/'      CONTOUR INTERVAL =
     1',F12.4)
      WRITE(6,3)
3     FORMAT('0 KEY TO CONTOUR VALUES'//)
      DO 5 I=KMIN,KMAX
      J=11-I
      P=REFC-J*CINT
      Q=P+CINT
      IF(I.EQ.1) GO TO 7
      IF(I.EQ.20) GO TO 8
      WRITE(6,9) (SYMB(I),J=1,6),P,Q
      GO TO 5
7     WRITE(6,10) (SYMB(I),J=1,6),P
10    FORMAT(' ',6A1,2X,F10.4,'AND BELOW')
      GO TO 5
8     WRITE(6,11) (SYMB(I),J=1,6),P
11    FORMAT(' ',6A1,2X,F10.4,'AND ABOVE')
9     FORMAT(' ',6A1,2X,F10.4,' TO ',F10.4)
5     CONTINUE
      WRITE(6,13) UMIN,VMAX,UMAX,VMAX,UMIN,VMIN,UMAX,VMIN
13    FORMAT('0 COORDINATES OF MAP CORNERS ARE:'/
     15X,' TOP LEFT',5X,2F12.4/
     15X,' TOP RIGHT    ',2F12.4/
     15X,' BOTTOM LEFT  ',2F12.4/
     15X,' BOTTOM RIGHT ',2F12.4)
66    CONTINUE
      RETURN
      END
```

```
      FUNCTION PROB(PESTS,WORK,N,NORD,U,V,JZ)
*** COMPUTES VALUE ON PROBABILITY SURFACE AT EACH POINT *********
      DIMENSION PESTS(N),WORK(N)
      QZ1=PESTS(1)+(PESTS(2)*U)+(PESTS(3)*V)
      QZ2=EXP(QZ1)
      ZZ=1.0+QZ2
      IF(JZ.EQ.2) GO TO 8
      PROB=QZ2/ZZ
      GO TO 9
      PROB=1.0/ZZ
      CONTINUE
      IF(NORD.LT.2) RETURN
      WORK(2)=U
      WORK(3)=V
      DO 1 M=2,NORD
      L1=M-1
      L2=M*(M+1)/2
      DO 2 I=1,M
      L2=L2+1
      NW200=L2-M
      WORK(L2)=WORK(NW200)*U
      QZ1=QZ1+PESTS(L2)*WORK(L2)
      NW201=L2+1
      NW202=L2-M
      WORK(NW201)=WORK(NW202)*V
      QZ1=QZ1+PESTS(NW201)*WORK(NW201)
      QZ2=EXP(QZ1)
      ZZ=1.0+QZ2
      IF(JZ.EQ.2) GO TO 4
      PROB=QZ2/ZZ
      GO TO 1
      PROB=1.0/ZZ
      CONTINUE
      RETURN
      END
```

Appendix 1(b) A Fortran program for the three response category case

The program presented in Table 7 basically follows the same structure as that in Table 6 and discussed in (a) above. The reader should therefore ensure that he has understood that discussion and should then take note of the following alterations.

a) Input alterations The first card now specifies the highest order of surface the user wishes to fit (IORD); the number of respondents in the sample (NSAMPL); the number of respondents in the sample who have chosen response category one, in terms of the example in III (iv) the number who have claimed to shop regularly at the hypermarket (N1); and the number who have chosen response category two, in terms of the example in III (iv) the number who have claimed to shop occasionally at the hypermarket (N2B). The input order of the set of cards which follow and which give the geographical co-ordinates now demands that all respondents choosing response category 1 are read in first, followed by all respondents choosing response category 2, followed by all respondents choosing response category 3 (this is the arbitrarily chosen response category whose parameters have been set to zero, see III (i)). In the example in III (iv) Table 4 must therefore be reorganised so that all housewives claiming to shop regularly at the hypermarket are read in first, followed by all housewives claiming to shop occasionally at the hypermarket, followed by all housewives claiming never to shop at the hypermarket.

b) Output alterations The output follows the structure outlined in (a) above and illustrated in Figures 6, 7, 8, 10 and Table 4. The user should be aware however, of the manner in which the parameter estimates, standard errors and parameter estimate/standard error ratios are printed in the multiple response case. Using 1st order probability surface models of the 3 response category kind as an example, the parameter estimates, standard errors and parameter estimate/standard error ratios are printed as follows.

PARAMETER ESTIMATES	STANDARD ERRORS	PARAMETER EST/STANDARD ERROR
-0.5533	0.1960	-2.8225
0.0131	..	..
-0.0454	..	..
-0.2395	..	..
0.0112	..	..
-0.0699	..	..

The first three parameter estimates are α_0, β_{10}, β_{20}; the second three are α_1, β_{11}, β_{21} (see III (i) for clarification). That is to say, the full set of parameter estimates associated with category 1 are printed prior to the full set of parameter estimates associated with response category 2.

c) Size restriction alterations The maximum number of respondents in the sample (NSAMPL) allowed remains the same as in the earlier program, i.e. 250. The maximum number who choose response category one (N1) allowed, is set at 100. The maximum number who choose response category two (N2B) allowed, is set at 100. The maximum number who choose response category three (NSAMPL-N2B-N1) allowed, is set at 150. These size restrictions, like those of the earlier program, are easily changed however, and in Appendix 2(b) the changes required in the program when NSAMPL=AT1, N1=AT2, N2B=AT3 and (NSAMPL-N2B-N1)=AT4 are given, thus allowing the user to modify the size restrictions of the program to suit his own needs.

Table 7

```
      ****** PROBABILITY SURFACE PROGRAMME FOR A 3 CATEGORY SITUATION.
      WRITTEN BY NEIL WRIGLEY, DEPT  OF GEOGRAPHY, UNIVERSITY OF BRISTOL

      COMMON/ONE/A(250,15),A2(100,15),A3(100,15),A1(150,15),NSAMPL,N1,N2
     1,N1A,N2A,N2B,N4B,K,K1,KA2

      ***** READ IN IORD=HIGHEST SURFACE ORDER TO BE FITTED, 1,2,3, OR 4
      NSAMPL=NUMBER OF SAMPLE LOCALITIES, N1=NUMBER OF LOCALITIES WHERE
      RESPONSE CATEGORY ONE IS RECORDED, N2B=NUMBER OF LOCALITIES WHERE
      RESPONSE CATEGORY TWO IS RECORDED ****************************
      **** DATA CARD: COL 1=IORD, COL 2=BLANK, COL 3-5=NSAMPL, COL 6=
      BLANK, COL 7-9=N1, COL 10=BLANK, COL 11-13=N2B *****************
      READ(5,1) IORD,NSAMPL,N1,N2B
1     FORMAT(I1,1X,I3,1X,I3,1X,I3)

      **** READ IN VERT=LENGTH OF PROBABILITY SURFACE MAP REQUIRED IN
      INCHES, WID=WIDTH IN INCHES, REFC=VALUE OF REFERENCE CONTOUR, CINT
      =CONTOUR INTERVAL, UMAX,UMIN=MAXIMUM AND MINIMUM ALONG HORIZONTAL
      AXIS, VMAX,VMIN=MAXIMUM AND MINIMUM ALONG VERTICAL AXIS *********
      *** DATA CARD: COLS 1-7=VERT, COLS 8-14=WID, COLS 15-19=REFC,
      COLS 20-24=CINT, COLS 25-31=UMAX, COLS 32-38=UMIN, COLS 39-45=VMAX
      COLS 46-52=VMIN ********************************************
      READ(5,18) VERT,WID,REFC,CINT,UMAX,UMIN,VMAX,VMIN
8     FORMAT(2F7.4,2F5.3,4F7.4)
      NL=INT(VERT*8.0+0.5)
      NC=INT(WID*10.0+0.5)
      UAX=(UMAX-UMIN)/FLOAT(NC-1)
      VAX=(VMAX-VMIN)/FLOAT(NL-1)
      UAX=UAX+0.000001*UAX
      VAX=VAX+0.000001*VAX

      **** READ IN GEOGRAPHICAL CO-ORDINATES OF EACH SAMPLE LOCALITY,
      U(I)=HORIZONTAL CO-ORDINATE, V(I)=VERTICAL CO-ORDINATE, IN FORMAT
      (F8.4,1X,F8.4) (N.B. IN THE PROGRAMME THESE CO-ORDINATES ARE
      STORED IN MATRIX A) *** ORDER OF CARDS-ALL SAMPLE LOCALITIES WHERE
      RESPONSE CATEGORY ONE IS RECORDED, FOLLOWED BY ALL LOCALITIES
      WHERE RESPONSE CATEGORY TWO IS RECORDED, FOLLOWED BY ALL
      LOCALITIES WHERE REMAINING RESPONSE CATEGORY IS RECORDED ** CREATE
      CONSTANT TERM *** PRINT SAMPLE LOCALITITIES *******************
      WRITE(6,11)
1     FORMAT('0       U             V')
      DO 3 I=1,NSAMPL
      A(I,1)=1.0
      READ(5,4) A(I,2),A(I,3)
      WRITE(6,44) (A(I,J),J=2,3)
      FORMAT(F8.4,1X,F8.4)
4     FORMAT(' ',2F9.4)

      ***** START OF MAIN LOOP ***********************************
      ID=0
      CALL SURF(ID,NL,NC,UAX,VAX,UMAX,UMIN,VMAX,VMIN,REFC,CINT)
      DO 7 I=1,IORD
      ID=ID+1
      CALL SURF(ID,NL,NC,UAX,VAX,UMAX,UMIN,VMAX,VMIN,REFC,CINT)
      CONTINUE
      STOP
      END


      SUBROUTINE SURF(IORD,NL,NC,UAX,VAX,UMAX,UMIN,VMAX,VMIN,REFC,CINT)
      COMMON/ONE/A(250,15),A2(100,15),A3(100,15),A1(150,15),NSAMPL,N1,N2
     1,N1A,N2A,N2B,N4B,K,K1,KA2
      DOUBLE PRECISION BZ1,DBLE
      DOUBLE PRECISION X,F,G,STEP,ACC,W,FMINUS,STER,TSTAT
      DIMENSION C10(250),C11(250),C13(250)
      DIMENSION PROB1(250),PROB2(250),PROB4(250),XZ1(15,1),XZ2(15,1)
      DIMENSION QZA(250,1),QZB(250,1),W(1981),PESTS2(30)
      DIMENSION X(30),G(30),TSTAT(30),STER(30),WORK(30),PESTS1(30)

      *** THIS IS THE MAIN PART OF THE PROGRAMME. IT CALCULATES THE
      PARAMETER ESTIMATES OF THE PROBABILITY SURFACE MODEL USING MAXIMUM
      LIKELIHOOD ESTIMATION. IT USES POWELL'S HYBRID STEEPEST DESCENT
      AND GENERALISED NEWTON METHOD *****************************

      **** FOLLOWING ARBITRARILY SETS THE ITERATION INSTRUCTIONS *****
      MAXFUN=2000
```

```
      IPRINT=20
      IF(IORD.EQ.0) GO TO 9650
      BZ=FLOAT(IORD)
      BZ1=DBLE(BZ*0.04)
      STEP=BZ1+0.03D0
      IF(IORD.EQ.1) ACC=0.16D0
      IF(IORD.EQ.2) ACC=0.32D0
      IF(IORD.EQ.3) ACC=0.54D0
      IF(IORD.EQ.4) ACC=0.80D0
      GO TO 9651
 9650 STEP=0.06D0
      ACC=0.08D0
C
C     ******* CALCULATE NUMBER OF PARAMETERS *************************
 9651 K=((IORD+1)*(IORD+2))/2
      N=K*2
      N1A=N1+1
      N2=N2B+N1
      N2A=N2+1
      N4B=NSAMPL-N2
      K1=K+1
      KA2=K*2
      KIZ=K1+1
      J81=0
      J92=0
C
C     ****** GENERATE ARBITRARY INITIAL PARAMETER ESTIMATES *********
      IF(IORD.EQ.0) GO TO 9652
      X(1)=1.0D0
      DO 2 I=2,K
 2    X(I)=0.1D0
      X(K1)=1.0D0
      DO 703 I=KIZ,N
 703  X(I)=0.1D0
      GO TO 9655
 9652 DO 9653 I=1,N
 9653 X(I)=0.1D0
      GO TO 9654
C
C     ******* GENERATE POLYNOMIAL TERMS *****************************
 9655 IF(IORD.EQ.1) GO TO 12
      KD=IORD-1
      KDA=((KD+1)*(KD+2))/2
      KDB=KDA+1
      KE=IORD-2
      KEA=((KE+1)*(KE+2))/2
      KEB=KEA+1
      DO 15 J=KEB,KDA
      DO 14 I=1,NSAMPL
 14   A(I,KDB)=A(I,J)*A(I,2)
      KDB=KDB+1
 15   CONTINUE
      DO 16 I=1,NSAMPL
      A(I,KDB)=A(I,KDA)*A(I,3)
 16   CONTINUE
 12   CONTINUE
C
C     ***** CALCULATE PARAMETER ESTIMATES ***************************
 9654 DO 5 I=1,N1
      DO 5 J=1,K
 5    A2(I,J)=A(I,J)
      DO 6 I=N1A,N2
      J81=J81+1
      DO 6 J=1,K
 6    A3(J81,J)=A(I,J)
      DO 8 I=N2A,NSAMPL
      J92=J92+1
      DO 8 J=1,K
 8    A1(J92,J)=A(I,J)
      IF(IORD.EQ.0) WRITE(6,9660)
 9660 FORMAT(/////,1X,'FOLLOWING IS THE MAXIMIZED LOG LIKELIHOOD VA
     1OR INTERCEPT ONLY MODEL')
      IF(IORD.EQ.0) GO TO 9661
      WRITE(6,6200) IORD
 6200 FORMAT(////////,1X,'PROBABILITY SURFACE MODEL OF ORDER ',I1,'
     1FITTED AND SURFACES DRAWN')
 9661 CONTINUE
      CALL VA06AD(N,X,F,G,STEP,ACC,MAXFUN,IPRINT,W)
      FMINUS=-1.0D0*F
      WRITE(6,369) FMINUS
 369  FORMAT(//,1X,'MAXIMIZED LOG LIKELIHOOD VALUE = ',F12.5)
```

```
      IF(IORD.EQ.0) RETURN
      NW12=0.5*N*(N+1)
      NW13=NW12+1
      NLESS=N-1
      N70=NW13
      N200=NW13+NLESS
      DO 6001 K2004=1,N
      N71=NLESS-K2004
      STER(K2004)=DSQRT(W(N70))
      N70=N200+1
6001  N200=N70+N71
      WRITE(6,2008)
2008  FORMAT(//,1X,'PARAMETER ESTIMATES  STANDARD ERRORS  PARAMETER EST/
     1STANDARD ERROR')
      DO 2007 I=1,N
      TSTAT(I)=X(I)/STER(I)
 007  WRITE(6,2009) X(I),STER(I),TSTAT(I)
 009  FORMAT(1X,F15.7,F18.7,F20.7)
      WRITE(6,6009)
 009  FORMAT(//,1X,'VARIANCE-COVARIANCE MATRIX OF PARAMETER ESTIMATES')
 001  FORMAT('-',3X,'COLUMN',I6,9I11)
 003  FORMAT(//,1X,'ROW'/)
 002  FORMAT(' ','ROW ',I3,3X,10F11.5)
      IBEGIN=NW13-1-N
      DO 7002 I=1,N,10
      ILESS1=I-1
      IEND=I+9
      IF(IEND.GT.N) IEND=N
      WRITE(6,8001) (ICOL,ICOL=I,IEND)
      WRITE(6,8003)
      DO 7002 J=I,N
      MAX=J
      IF(MAX.GT.IEND) MAX=IEND
      IBASE=IBEGIN+J
      DO 7001 K17=I,MAX
      INDEX=IBASE+K17*N-(K17*K17-K17)/2
      L=K17-ILESS1
 001  PROB1(L)=SNGL(W(INDEX))
      MAX=MAX-ILESS1
 002  WRITE(6,8002) J,(PROB1(L),L=1,MAX)
      JZ2=0
      DO 630 I=1,K
 30   XZ1(I,1)=SNGL(X(I))
      DO 631 I=K1,KA2
      JZ2=JZ2+1
 31   XZ2(JZ2,1)=SNGL(X(I))

      ***** COMPUTE AND PRINT ESTIMATED PROBABILITIES AND RESIDUALS ***
      CALL MC01AS(A,XZ1,QZA,NSAMPL,K,1,250,15,250)
      CALL MC01AS(A,XZ2,QZB,NSAMPL,K,1,250,15,250)
      DO 633 I=1,NSAMPL
      QZA(I,1)=EXP(QZA(I,1))
      QZB(I,1)=EXP(QZB(I,1))
      ZZ=1.0+QZA(I,1)+QZB(I,1)
      PROB1(I)=QZA(I,1)/ZZ
      PROB2(I)=QZB(I,1)/ZZ
 3    PROB4(I)=1.0/ZZ
      DO 8910 I=1,NSAMPL
      IF(PROB1(I).EQ.1.0) PROB1(I)=0.9999999
      IF(PROB1(I).EQ.0.0) PROB1(I)=0.0000001
      IF(PROB2(I).EQ.1.0) PROB2(I)=0.9999999
      IF(PROB2(I).EQ.0.0) PROB2(I)=0.0000001
      IF(PROB4(I).EQ.1.0) PROB4(I)=0.9999999
      IF(PROB4(I).EQ.0.0) PROB4(I)=0.0000001
      C10(I)=SQRT(PROB1(I)*(1.0-PROB1(I)))
      C11(I)=SQRT(PROB2(I)*(1.0-PROB2(I)))
      C13(I)=SQRT(PROB4(I)*(1.0-PROB4(I)))
 10   CONTINUE
      WRITE(6,8901)
 01   FORMAT(///,1X,'FOR EACH SAMPLE LOCALITY'/'PREDICTED PROBABILITES
     1                RAW RESIDUALS                             STANDARDISED RES
     1IDUALS')
      WRITE(6,8930)
 30   FORMAT(/,1X,'RESP CAT 1  RESP CAT 2  RESP CAT 3  RESP CAT 1  RESP
     1CAT 2  RESP CAT 3  RESP CAT 1  RESP CAT 2  RESP CAT 3')
      DO 8920 I=1,N1
      RAWR1=(1.0-PROB1(I))
      RAWR2=-PROB2(I)
      RAWR4=-PROB4(I)
      RES1=RAWR1/C10(I)
      RES2=RAWR2/C11(I)
      RES4=RAWR4/C13(I)
```

```
8920 WRITE(6,8902) PROB1(I),PROB2(I),PROB4(I),RAWR1,RAWR2,RAWR4,RES1,
    1S2,RES4
     DO 8921 I=N1A,N2
     RAWR1=-PROB1(I)
     RAWR2=(1.0-PROB2(I))
     RAWR4=-PROB4(I)
     RES1=RAWR1/C10(I)
     RES2=RAWR2/C11(I)
     RES4=RAWR4/C13(I)
8921 WRITE(6,8902) PROB1(I),PROB2(I),PROB4(I),RAWR1,RAWR2,RAWR4,RES1,
    1S2,RES4
     DO 8922 I=N2A,NSAMPL
     RAWR1=-PROB1(I)
     RAWR2=-PROB2(I)
     RAWR4=(1.0-PROB4(I))
     RES1=RAWR1/C10(I)
     RES2=RAWR2/C11(I)
     RES4=RAWR4/C13(I)
8922 WRITE(6,8902) PROB1(I),PROB2(I),PROB4(I),RAWR1,RAWR2,RAWR4,RES1
    1S2,RES4
8902 FORMAT(' ',F8.5,4X,F8.5,4X,F8.5,3X,6(F9.5,3X))
     DO 243 I=1,K
     PESTS1(I)=XZ1(I,1)
243  PESTS2(I)=XZ2(I,1)
     CALL PMAP(NL,NC,PESTS1,PESTS2,WORK,REFC,CINT,UAX,VAX,N,IORD,VMA
    1MIN,UMAX,VMIN)
     RETURN
     END




     SUBROUTINE CALCFG(N,X,F,G)
     DOUBLE PRECISION X,G,U12,U13,U18,F,DBLE
     COMMON/ONE/A(250,15),A2(100,15),A3(100,15),A1(150,15),NSAMPL,N1
    1,N1A,N2A,N2B,N4B,K,K1,KA2
     DIMENSION X(30),G(30)
     DIMENSION XN1(15,1),XN2(15,1),Q1(100,1),Q2(100,1)
     DIMENSION U6(250,1),U7(250,1),U9(250,1)
     DIMENSION ROW1(1,250),ROW2(1,100),ROW3(1,100)
     DIMENSION U11(250,15),U12(1,15),U13(1,15),U18(1,15)
     J2=0
     DO 66 I=1,K
66   XN1(I,1)=SNGL(X(I))
     DO 67 I=K1,KA2
     J2=J2+1
67   XN2(J2,1)=SNGL(X(I))
     U1=0.0
     U2=0.0
     Z1=0.0
     CALL MC01AS(A2,XN1,Q1,N1,K,1,100,15,100)
     CALL MC01AS(A3,XN2,Q2,N2B,K,1,100,15,100)
     DO 169 I=1,N1
169  U1=U1+Q1(I,1)
     DO 70 I=1,N2B
70   U2=U2+Q2(I,1)
     Z=U1+U2
     CALL MC01AS(A,XN1,U6,NSAMPL,K,1,250,15,250)
     DO 87 I=1,NSAMPL
87   U6(I,1)=EXP(U6(I,1))
     CALL MC01AS(A,XN2,U7,NSAMPL,K,1,250,15,250)
     DO 88 I=1,NSAMPL
88   U7(I,1)=EXP(U7(I,1))
     DO 90 I=1,NSAMPL
     U9(I,1)=U6(I,1)+U7(I,1)
90   U9(I,1)=1.0+U9(I,1)
     DO 91 I=1,NSAMPL
91   Z1=Z1+ALOG(U9(I,1))
     FA=-1.0*(Z-Z1)
     F=DBLE(FA)
C    ***** -F IS THE LOG LIKELIHOOD VALUE ************************
C
C    *** CALCULATE 1ST DERIVATIVES OF LOG LIKELIHOOD FUNCTION ****
     DO 44 I=1,NSAMPL
44   ROW1(1,I)=1.0
     DO 45 I=1,N1
45   ROW2(1,I)=1.0
     DO 46 I=1,N2B
46   ROW3(1,I)=1.0
     DO 94 I=1,NSAMPL
94   U6(I,1)=U6(I,1)/U9(I,1)
```

```
      DO 92  J=1,K
      DO 92  I=1,NSAMPL
92    U11(I,J)=A(I,J)*U6(I,1)
      CALL MATNW(ROW1,U11,U12,1,NSAMPL,K,1,250,1)
      CALL MATNW(ROW2,A2,U13,1,N1,K,1,100,1)
      DO 93  J=1,K
93    U13(1,J)=U13(1,J)-U12(1,J)
      DO 95  I=1,NSAMPL
5     U7(I,1)=U7(I,1)/U9(I,1)
      DO 96  J=1,K
      DO 96  I=1,NSAMPL
6     U11(I,J)=A(I,J)*U7(I,1)
      CALL MATNW(ROW1,U11,U12,1,NSAMPL,K,1,250,1)
      CALL MATNW(ROW3,A3,U18,1,N2B,K,1,100,1)
      DO 97  J=1,K
7     U18(1,J)=U18(1,J)-U12(1,J)
      DO 101  I=1,K
 01   G(I)=U13(1,I)
      DO 102  I=1,K
      J4=I+K
 02   G(J4)=U18(1,I)
      DO 104  I=1,N
 04   G(I)=-G(I)
      ****  G IS THE VECTOR OF 1ST DERIVATIVES ***********************
      RETURN
      END



      SUBROUTINE PMAP(NL,NC,PESTS1,PESTS2,WORK,REFC,CINT,UAX,VAX,N,IORD,
     1VMAX,UMIN,UMAX,VMIN)
      DIMENSION PESTS1(N),PESTS2(N),WORK(N),SLINE(120),SYMB(23)
      DATA SYMB/1H1,1H.,1H2,1H.,1H3,1H.,1H4,1H.,1H5,1H.,1H\,1H.,1H6,1H.,
     11H7,1H.,1H8,1H.,1H9,1H.,1H0,1HI,1H-/
      FORMAT('1        PROBABILITY SURFACE OF ORDER ',I8/' PART',I4//)
      FORMAT(' ',120A1)
      DO 66  JZ=1,3
      KMAX=0
      KMIN=20
      NO2=110
      NSTR=NC/110
      LEFT=NC-110*NSTR
      IF(LEFT.GT.0) NSTR=NSTR+1
      DO 96  KOUNT=1,NSTR
      IF(KOUNT.EQ.NSTR.AND.LEFT.GT.0) NO2=LEFT
      WRITE(6,1)  IORD,KOUNT
      NO22=NO2+2
      WRITE(6,2)  (SYMB(23),I=1,NO22)
      V=VMAX+VAX
      DO 12  L=1,NL
      V=V-VAX
      U=UMIN-UAX+(KOUNT-1)*110*UAX
      DO 6  I=1,NO2
      U=U+UAX
      PR=PROB(PESTS1,PESTS2,WORK,N,IORD,U,V,JZ)
      KCC=11+INT((PR-REFC)/CINT)
      IF(PR.LT.REFC) KCC=KCC-1
      IF(KCC.LT.1) KCC=1
      IF(KCC.GT.20) KCC=20
      IF(KCC.GT.KMAX) KMAX=KCC
      IF(KCC.LT.KMIN) KMIN=KCC
      SLINE(I)=SYMB(KCC)
      WRITE(6,2)  SYMB(22),(SLINE(I),I=1,NO2),SYMB(22)
      WRITE(6,2)  (SYMB(23),I=1,NO22)
      CONTINUE
      WRITE(6,4)  REFC,CINT
      FORMAT('0     REFERENCE CONTOUR = ',F11.4/'      CONTOUR INTERVAL =
     1',F12.4)
      WRITE(6,3)
      FORMAT('0 KEY TO CONTOUR VALUES'//)
      DO 5  I=KMIN,KMAX
      J=11-I
      P=REFC-J*CINT
      Q=P+CINT
      IF(I.EQ.1) GO TO 7
      IF(I.EQ.20) GO TO 8
      WRITE(6,9)  (SYMB(I),J=1,6),P,Q
      GO TO 5
      WRITE(6,10)  (SYMB(I),J=1,6),P
```

```
 10     FORMAT(' ',6A1,2X,F10.4,'AND BELOW')
        GO TO 5
 8      WRITE(6,11) (SYMB(I),J=1,6),P
 11     FORMAT(' ',6A1,2X,F10.4,'AND ABOVE')
 9      FORMAT(' ',6A1,2X,F10.4,' TO ',F10.4)
 5      CONTINUE
        WRITE(6,13) UMIN,VMAX,UMAX,VMAX,UMIN,VMIN,UMAX,VMIN
 13     FORMAT('0 COORDINATES OF MAP CORNERS ARE:'/
       15X,' TOP LEFT',5X,2F12.4/
       15X,' TOP RIGHT     ',2F12.4/
       15X,' BOTTOM LEFT  ',2F12.4/
       15X,' BOTTOM RIGHT ',2F12.4)
 66     CONTINUE
        RETURN
        END




        FUNCTION PROB(PESTS1,PESTS2,WORK,N,NORD,U,V,JZ)
 C      *** COMPUTES VALUE ON PROBABILITY SURFACE AT EACH POINT *******
        DIMENSION PESTS1(N),PESTS2(N),WORK(N)
        QZ1A=PESTS1(1)+(PESTS1(2)*U)+(PESTS1(3)*V)
        QZ1B=PESTS2(1)+(PESTS2(2)*U)+(PESTS2(3)*V)
        QZ2A=EXP(QZ1A)
        QZ2B=EXP(QZ1B)
        ZZ=1.0+QZ2A+QZ2B
        IF(JZ.EQ.2) GO TO 8
        IF(JZ.EQ.3) GO TO 9
        PROB=QZ2A/ZZ
        GO TO 10
 8      PROB=QZ2B/ZZ
        GO TO 10
 9      PROB=1.0/ZZ
 10     CONTINUE
        IF(NORD.LT.2) RETURN
        WORK(2)=U
        WORK(3)=V
        DO 1 M=2,NORD
        L1=M-1
        L2=M*(M+1)/2
        DO 2 I=1,M
        L2=L2+1
        NW200=L2-M
        WORK(L2)=WORK(NW200)*U
        QZ1A=QZ1A+PESTS1(L2)*WORK(L2)
 2      QZ1B=QZ1B+PESTS2(L2)*WORK(L2)
        NW201=L2+1
        NW202=L2-M
        WORK(NW201)=WORK(NW202)*V
        QZ1A=QZ1A+PESTS1(NW201)*WORK(NW201)
        QZ1B=QZ1B+PESTS2(NW201)*WORK(NW201)
        QZ2A=EXP(QZ1A)
        QZ2B=EXP(QZ1B)
        ZZ=1.0+QZ2A+QZ2B
        IF(JZ.EQ.2) GO TO 4
        IF(JZ.EQ.3) GO TO 5
        PROB=QZ2A/ZZ
        GO TO 1
 4      PROB=QZ2B/ZZ
        GO TO 1
 5      PROB=1.0/ZZ
 1      CONTINUE
        RETURN
        END
```

Appendix 1(c) A Fortran program for the four response category case

The program presented in Table 8 follows the same structure as that discussed in (b) above and the minor alterations from the program in Table 7 are explained in the program itself. Appendix 2(c) contains the changes necessary to allow the user to modify the size restrictions of the program to suit his own needs. The program, Table 8, begins on page 65.

Appendix 1(d) The common subroutines

The programs listed in Tables 6, 7 and 8 utilise the common subroutines presented in Table 9. Subroutine VA06AD is taken from the Harwell Algorithm Package and was written by M.J.D. Powell. The listing given below incorporates some minor changes by the author but it is essentially in its origina form. It is reproduced here with the kind permission of the Controllers of the Harwell Algorithms Package.

APPENDIX 2 Alterations to size restrictions of computer programs

Appendix 2(a)

The following are the card changes necessary in the program listed in Table 6 when NSAMPL=AT1, N1=AT2 and (NSAMPL-N1)=AT3.

In main part of program:

```
COMMON/ONE/A(AT1,15),A2(AT2,15),A1(AT3,15),NSAMPL,N1,N1A,K,N4B
```

In subroutine SURF:

```
COMMON/ONE/A(AT1,15),A2(AT2,15),A1(AT3,15),NSAMPL,N1,N1A,K,N4B
DIMENSION C10(AT1),C13(AT1)
DIMENSION PROB1(AT1),PROB4(AT1),QZA(AT1,1),XZ1(15,1)
CALL MC01AS(A,XZ1,QZA,NSAMPL,K,1,AT1,15,AT1)
```

In subroutine CALCFG:

```
COMMON/ONE/A(AT1,15),A2(AT2,15),A1(AT3,15),NSAMPL,N1,N1A,K,N4B
DIMENSION XN1(15,1),Q1(AT2,1),U6(AT1,1),U9(AT1,1),ROW1(1,AT1)
DIMENSION ROW2(1,AT2),U11(AT1,15),U12(1,15),U13(1,15)
CALL MC01AS(A2,XN1,Q1,N1,K,1,AT2,15,AT2)
CALL MC01AS(A,XN1,U6,NSAMPL,K,1,AT1,15,AT1)
CALL MATNW(ROW1,U11,U12,1,NSAMPL,K,1,AT1,1)
CALL MATNW(ROW2,A2,U13,1,N1,K,1,AT2,1)
```

Appendix 2(b)

The following are the card changes necessary in the program listed in Table 7 when NSAMPL=AT1, N1=AT2, N2B=AT3 and (NSAMPL-N2B-N1)=AT4.

In main part of program:

```
COMMON/ONE/A(AT1,15),A2(AT2,15),A3(AT3,15),A1(AT4,15),NSAMPL,N1,N2
```

In subroutine SURF:

```
COMMON/ONE/A(AT1,15),A2(AT2,15),A3(AT3,15),A1(AT4,15),NSAMPL,N1,N2
DIMENSION C10(AT1),C11(AT1),C13(AT1)
DIMENSION PROB1(AT1),PROB2(AT1),PROB4(AT1),XZ1(15,1),XZ2(15,1)
```

```
DIMENSION QZA(AT1,1),QZB(AT1,1),W(1981),PESTS2(30)
CALL MC01AS(A,XZ1,QZA,NSAMPL,K,1,AT1,15,AT1)
CALL MC01AS(A,XZ2,QZB,NSAMPL,K,1,AT1,15,AT1)
```

In subroutine CALCFG:

```
COMMON/ONE/A(AT1,15),A2(AT2,15),A3(AT3,15),A1(AT4,15),NSAMPL,N1,N2
DIMENSION XN1(15,1),XN2(15,1),Q1(AT2,1),Q2(AT3,1)
DIMENSION U6(AT1,1),U7(AT1,1),U9(AT1,1)
DIMENSION ROW1(1,AT1),ROW2(1,AT2),ROW3(1,AT3)
DIMENSION U11(AT1,15),U12(1,15),U13(1,15),U18(1,15)
CALL MC01AS(A2,XN1,Q1,N1,K,1,AT2,15,AT2)
CALL MC01AS(A3,XN2,Q2,N2B,K,1,AT3,15,AT3)
CALL MC01AS(A,XN1,U6,NSAMPL,K,1,AT1,15,AT1)
CALL MC01AS(A,XN2,U7,NSAMPL,K,1,AT1,15,AT1)
CALL MATNW(ROW1,U11,U12,1,NSAMPL,K,1,AT1,1)
CALL MATNW(ROW2,A2,U13,1,N1,K,1,AT2,1)
CALL MATNW(ROW1,U11,U12,1,NSAMPL,K,1,AT1,1)
CALL MATNW(ROW3,A3,U18,1,N2B,K,1,AT3,1)
```

Appendix 2(c)

The following are the card changes necessary in the program listed in Table 8 when NSAMPL=AT1, N1=AT2, N2B=AT3, N3B=AT4 and (NSAMPL-N3B-N2B-N1) =AT5.

In main part of program:

```
COMMON/ONE/A(AT1,15),A2(AT2,15),A3(AT3,15),A4(AT4,15),A1(AT5,15),N
```

In subroutine SURF:

```
COMMON/ONE/A(AT1,15),A2(AT2,15),A3(AT3,15),A4(AT4,15),A1(AT5,15),N
DIMENSION C10(AT1),C11(AT1),C12(AT1),C13(AT1)
DIMENSION PROB1(AT1),PROB2(AT1),PROB3(AT1),PROB4(AT1)
DIMENSION QZA(AT1,1),QZB(AT1,1),QZC(AT1,1),TSTAT(45),STER(45)
CALL MC01AS(A,XZ1,QZA,NSAMPL,K,1,AT1,15,AT1)
CALL MC01AS(A,XZ2,QZB,NSAMPL,K,1,AT1,15,AT1)
CALL MC01AS(A,XZ3,QZC,NSAMPL,K,1,AT1,15,AT1)
```

In subroutine CALCFG:

```
COMMON/ONE/A(AT1,15),A2(AT2,15),A3(AT3,15),A4(AT4,15),A1(AT5,15),N
DIMENSION X(45),G(45),Q1(AT2,1),Q2(AT3,1),Q3(AT4,1)
DIMENSION U6(AT1,1),U7(AT1,1),U8(AT1,1),U9(AT1,1)
DIMENSION ROW1(1,AT1),ROW2(1,AT2),ROW3(1,AT3),ROW4(1,AT4)
DIMENSION U11(AT1,15),U12(1,15),U13(1,15),U18(1,15),U23(1,15)
CALL MC01AS(A2,XN1,Q1,N1,K,1,AT2,15,AT2)
CALL MC01AS(A3,XN2,Q2,N2B,K,1,AT3,15,AT3)
CALL MC01AS(A4,XN3,Q3,N3B,K,1,AT4,15,AT4)
CALL MC01AS(A,XN1,U6,NSAMPL,K,1,AT1,15,AT1)
CALL MC01AS(A,XN2,U7,NSAMPL,K,1,AT1,15,AT1)
CALL MC01AS(A,XN3,U8,NSAMPL,K,1,AT1,15,AT1)
CALL MATNW(ROW1,U11,U12,1,NSAMPL,K,1,AT1,1)
CALL MATNW(ROW2,A2,U13,1,N1,K,1,AT2,1)
CALL MATNW(ROW1,U11,U12,1,NSAMPL,K,1,AT1,1)
CALL MATNW(ROW3,A3,U18,1,N2B,K,1,AT3,1)
CALL MATNW(ROW1,U11,U12,1,NSAMPL,K,1,AT1,1)
CALL MATNW(ROW4,A4,U23,1,N3B,K,1,AT4,1)
```

able 8

```
***** PROBABILITY SURFACE PROGRAMME FOR A 4 CATEGORY SITUATION
WRITTEN BY NEIL WRIGLEY, DEPT. OF GEOGRAPHY, UNIVERSITY OF BRISTOL

COMMON/ONE/A(250,15),A2(100,15),A3(100,15),A4(100,15),A1(150,15),N
1SAMPL,N1,N2,N3,N1A,N2A,N3A,N2B,N3B,K,K1,KA2,K2

***** READ IN IORD=HIGHEST SURFACE ORDER TO BE FITTED, 1,2,3, OR 4
NSAMPL=NUMBER OF SAMPLE LOCALITIES, N1=NUMBER OF LOCALITIES WHERE
RESPONSE CATEGORY ONE IS RECORDED, N2B=NUMBER OF LOCALITIES WHERE
RESPONSE CATEGORY TWO IS RECORDED, N3B=NUMBER OF LOCALITIES WHERE
RESPONSE CATEGORY THREE IS RECORDED *****************************
**** DATA CARD: COL 1=IORD, COL 2=BLANK, COL 3-5=NSAMPL, COL 6=
BLANK, COL 7-9=N1, COL 10=BLANK, COL 11-13=N2B, COL 14=BLANK, COL
15-17=N3B ******************************************************
READ(5,1) IORD,NSAMPL,N1,N2B,N3B
FORMAT(I1,1X,I3,1X,I3,1X,I3,1X,I3)

**** READ IN VERT=LENGTH OF PROBABILITY SURFACE MAP REQUIRED IN
INCHES, WID=WIDTH IN INCHES, REFC=VALUE OF REFERENCE CONTOUR, CINT
=CONTOUR INTERVAL, UMAX,UMIN=MAXIMUM AND MINIMUM ALONG HORIZONTAL
AXIS, VMAX,VMIN=MAXIMUM AND MINIMUM ALONG VERTICAL AXIS *********
*** DATA CARD, COLS 1-7=VERT, COLS 8-14=WID, COLS 15-19=REFC,
COLS 20-24=CINT, COLS 25-31=UMAX, COLS 32-38=UMIN, COLS 39-45=VMAX
COLS 46-52=VMIN ************************************************
READ(5,10) VERT,WID,REFC,CINT,UMAX,UMIN,VMAX,VMIN
FORMAT(2F7.4,2F5.3,4F7.4)
NL=INT(VERT*8.0+0.5)
NC=INT(WID*10.0+0.5)
UAX=(UMAX-UMIN)/FLOAT(NC-1)
VAX=(VMAX-VMIN)/FLOAT(NL-1)
UAX=UAX+0.000001*UAX
VAX=VAX+0.000001*VAX

**** READ IN GEOGRAPHICAL CO-ORDINATES OF EACH SAMPLE LOCALITY,
U(I)=HORIZONTAL CO-ORDINATE, V(I)=VERTICAL CO-ORDINATE, IN FORMAT
(F8.4,1X,F8.4) (N.B. IN THE PROGRAMME THESE CO-ORDINATES ARE
STORED IN MATRIX A) *** ORDER OF CARDS-ALL SAMPLE LOCALITIES WHERE
RESPONSE CATEGORY ONE IS RECORDED, FOLLOWED BY ALL LOCALITIES
WHERE RESPONSE CATEGORY TWO IS RECORDED, FOLLOWED BY ALL
LOCALITIES WHERE RESPONSE CATEGORY THREE IS RECORDED, FOLLOWED BY
ALL LOCALITIES WHERE REMAINING RESPONSE CATEGORY IS RECORDED *****
CREATE CONSTANT TERM *** PRINT SAMPLE LOCALITIES ****************
WRITE(6,11)
FORMAT('0      U          V')
DO 3 I=1,NSAMPL
A(I,1)=1.0
READ(5,4) A(I,2),A(I,3)
WRITE(6,44) (A(I,J),J=2,3)
FORMAT(F8.4,1X,F8.4)
FORMAT(' ',2F9.4)

***** START OF MAIN LOOP ****************************************
ID=0
CALL SURF(ID,NL,NC,UAX,VAX,UMAX,UMIN,VMAX,VMIN,REFC,CINT)
DO 7 I=1,IORD
ID=ID+1
CALL SURF(ID,NL,NC,UAX,VAX,UMAX,UMIN,VMAX,VMIN,REFC,CINT)
CONTINUE
STOP
END




SUBROUTINE SURF(IORD,NL,NC,UAX,VAX,UMAX,UMIN,VMAX,VMIN,REFC,CINT)
COMMON/ONE/A(250,15),A2(100,15),A3(100,15),A4(100,15),A1(150,15),N
1SAMPL,N1,N2,N3,N1A,N2A,N3A,N2B,N3B,K,K1,KA2,K2
DOUBLE PRECISION BZ1,DBLE
DOUBLE PRECISION X,F,G,STEP,ACC,W,FMINUS,STER,TSTAT
DIMENSION C10(250),C11(250),C12(250),C13(250)
DIMENSION PROB1(250),PROB2(250),PROB3(250),PROB4(250)
DIMENSION XZ1(15,1),XZ2(15,1),XZ3(15,1),W(4321),WORK(45)
DIMENSION QZA(250,1),QZB(250,1),QZC(250,1),TSTAT(45),STER(45)
DIMENSION X(45),G(45),PESTS1(45),PESTS2(45),PESTS3(45)

*** THIS IS THE MAIN PART OF THE PROGRAMME. IT CALCULATES THE
PARAMETER ESTIMATES OF THE PROBABILITY SURFACE MODEL USING MAXIMUM
LIKELIHOOD ESTIMATION. IT USES POWELL'S HYBRID STEEPEST DESCENT
```

```
C     AND GENERALISED NEWTON METHOD ****************************
C
C     **** FOLLOWING ARBITRARILY SETS THE ITERATION INSTRUCTIONS ***
      MAXFUN=2000
      IPRINT=20
      IF(IORD.EQ.0) GO TO 9650
      BZ=FLOAT(IORD)
      BZ1=DBLE(BZ*0.05)
      STEP=BZ1+0.03D0
      IF(IORD.EQ.1) ACC=0.24D0
      IF(IORD.EQ.2) ACC=0.48D0
      IF(IORD.EQ.3) ACC=0.71D0
      IF(IORD.EQ.4) ACC=1.20D0
      GO TO 9651
 9650 STEP=0.06D0
      ACC=0.08D0
C
C     ******* CALCULATE NUMBER OF PARAMETERS ***********************
 9651 K=((IORD+1)*(IORD+2))/2
      N=K*3
      N1A=N1+1
      N2=N2B+N1
      N2A=N2+1
      N3=N3B+N2
      N3A=N3+1
      K1=K+1
      KA2=K*2
      K2=KA2+1
      KIZ=K1+1
      KIZA=K2+1
      J81=0
      J91=0
      J92=0
C
C     ****** GENERATE ARBITRARY INITIAL PARAMETER ESTIMATES *******
      IF(IORD.EQ.0) GO TO 9652
      X(1)=1.0D0
      DO 2 I=2,K
 2    X(I)=0.1D0
      X(K1)=1.0D0
      DO 703 I=KIZ,KA2
 703  X(I)=0.1D0
      X(K2)=1.0D0
      DO 704 I=KIZA,N
 704  X(I)=0.1D0
      GO TO 9655
 9652 DO 9653 I=1,N
 9653 X(I)=0.1D0
      GO TO 9654
C
C     ******* GENERATE POLYNOMIAL TERMS ***************************
 9655 IF(IORD.EQ.1) GO TO 12
      KD=IORD-1
      KDA=((KD+1)*(KD+2))/2
      KDB=KDA+1
      KE=IORD-2
      KEA=((KE+1)*(KE+2))/2
      KEB=KEA+1
      DO 15 J=KEB,KDA
      DO 14 I=1,NSAMPL
 14   A(I,KDB)=A(I,J)*A(I,2)
      KDB=KDB+1
 15   CONTINUE
      DO 16 I=1,NSAMPL
      A(I,KDB)=A(I,KDA)*A(I,3)
 16   CONTINUE
 12   CONTINUE
C
C     ***** CALCULATE PARAMETER ESTIMATES *************************
 9654 DO 5 I=1,N1
      DO 5 J=1,K
 5    A2(I,J)=A(I,J)
      DO 6 I=N1A,N2
      J81=J81+1
      DO 6 J=1,K
 6    A3(J81,J)=A(I,J)
      DO 7 I=N2A,N3
      J91=J91+1
      DO 7 J=1,K
 7    A4(J91,J)=A(I,J)
      DO 8 I=N3A,NSAMPL
      J92=J92+1
```

```
      DO 8 J=1,K
8     A1(J92,J)=A(I,J)
      IF(IORD.EQ.0) WRITE(6,9660)
9660  FORMAT(/////,1X,'FOLLOWING IS THE MAXIMIZED LOG LIKELIHOOD VALUE F
     1OR INTERCEPT ONLY MODEL')
      IF(IORD.EQ.0) GO TO 9661
      WRITE(6,6200) IORD
,200  FORMAT(////////,1X,'PROBABILITY SURFACE MODEL OF ORDER ',I1,' NOW
     1FITTED AND SURFACES DRAWN')
9661  CONTINUE
      CALL VA06AD(N,X,F,G,STEP,ACC,MAXFUN,IPRINT,W)
      FMINUS=-1.0D0*F
      WRITE(6,369) FMINUS
369   FORMAT(//,1X,'MAXIMIZED LOG LIKELIHOOD VALUE = ',F12.5)
      IF(IORD.EQ.0) RETURN
      NW12=0.5*N*(N+1)
      NW13=NW12+1
      NLESS=N-1
      N70=NW13
      N200=NW13+NLESS
      DO 6001 K2004=1,N
      N71=NLESS-K2004
      STER(K2004)=DSQRT(W(N70))
      N70=N200+1
 001  N200=N70+N71
      WRITE(6,2008)
 008  FORMAT(//,1X,'PARAMETER ESTIMATES  STANDARD ERRORS  PARAMETER EST
     1STANDARD ERROR')
      DO 2007 I=1,N
      TSTAT(I)=X(I)/STER(I)
007   WRITE(6,2009) X(I),STER(I),TSTAT(I)
009   FORMAT(1X,F15.7,F18.7,F20.7)
      WRITE(6,6009)
009   FORMAT(//,1X,'VARIANCE-COVARIANCE MATRIX OF PARAMETER ESTIMATES')
001   FORMAT('-',3X,'COLUMN',I6,9I11)
003   FORMAT(//,1X,'ROW'/)
002   FORMAT(' ','ROW ',I3,3X,10F11.5)
      IBEGIN=NW13-1-N
      DO 7002 I=1,N,10
      ILESS1=I-1
      IEND=I+9
      IF(IEND.GT.N) IEND=N
      WRITE(6,8001) (ICOL,ICOL=I,IEND)
      WRITE(6,8003)
      DO 7002 J=I,N
      MAX=J
      IF(MAX.GT.IEND) MAX=IEND
      IBASE=IBEGIN+J
      DO 7001 K17=I,MAX
      INDEX=IBASE+K17*N-(K17*K17-K17)/2
      L=K17-ILESS1
01    PROB1(L)=SNGL(W(INDEX))
      MAX=MAX-ILESS1
02    WRITE(6,8002) J,(PROB1(L),L=1,MAX)
      JZ2=0
      JZ3=0
      DO 630 I=1,K
 0    XZ1(I,1)=SNGL(X(I))
      DO 631 I=K1,KA2
      JZ2=JZ2+1
 1    XZ2(JZ2,1)=SNGL(X(I))
      DO 632 I=K2,N
      JZ3=JZ3+1
 2    XZ3(JZ3,1)=SNGL(X(I))

      ***** COMPUTE AND PRINT ESTIMATED PROBABILITIES AND RESIDUALS ***
      CALL MC01AS(A,XZ1,QZA,NSAMPL,K,1,250,15,250)
      CALL MC01AS(A,XZ2,QZB,NSAMPL,K,1,250,15,250)
      CALL MC01AS(A,XZ3,QZC,NSAMPL,K,1,250,15,250)
      DO 633 I=1,NSAMPL
      QZA(I,1)=EXP(QZA(I,1))
      QZB(I,1)=EXP(QZB(I,1))
      QZC(I,1)=EXP(QZC(I,1))
      ZZ=1.0+QZA(I,1)+QZB(I,1)+QZC(I,1)
      PROB1(I)=QZA(I,1)/ZZ
      PROB2(I)=QZB(I,1)/ZZ
      PROB3(I)=QZC(I,1)/ZZ
3     PROB4(I)=1.0/ZZ
      DO 8910 I=1,NSAMPL
      IF(PROB1(I).EQ.1.0) PROB1(I)=0.9999999
      IF(PROB1(I).EQ.0.0) PROB1(I)=0.0000001
      IF(PROB2(I).EQ.1.0) PROB2(I)=0.9999999
      IF(PROB2(I).EQ.0.0) PROB2(I)=0.0000001
```

```
      IF(PROB3(I).EQ.1.0) PROB3(I)=0.9999999
      IF(PROB3(I).EQ.0.0) PROB3(I)=0.0000001
      IF(PROB4(I).EQ.1.0) PROB4(I)=0.9999999
      IF(PROB4(I).EQ.0.0) PROB4(I)=0.0000001
      C10(I)=SQRT(PROB1(I)*(1.0-PROB1(I)))
      C11(I)=SQRT(PROB2(I)*(1.0-PROB2(I)))
      C12(I)=SQRT(PROB3(I)*(1.0-PROB3(I)))
      C13(I)=SQRT(PROB4(I)*(1.0-PROB4(I)))
8910  CONTINUE
      WRITE(6,8901)
8901  FORMAT(///,1X,'FOR EACH SAMPLE LOCALITY'/'PREDICTED PROBABILITES
     1                    RAW RESIDUALS                              STANDAR
     1SED RESIDUALS')
      WRITE(6,8930)
8930  FORMAT(/,1X,'RSP CAT1  RSP CAT2  RSP CAT3  RSP CAT4  RSP CAT1  R
     1 CAT2  RSP CAT3  RSP CAT4  RSP CAT1  RSP CAT2  RSP CAT3  RSP CAT
     1)
      DO 8920 I=1,N1
      RAWR1=(1.0-PROB1(I))
      RAWR2=-PROB2(I)
      RAWR3=-PROB3(I)
      RAWR4=-PROB4(I)
      RES1=RAWR1/C10(I)
      RES2=RAWR2/C11(I)
      RES3=RAWR3/C12(I)
      RES4=RAWR4/C13(I)
8920  WRITE(6,8902) PROB1(I),PROB2(I),PROB3(I),PROB4(I),RAWR1,RAWR2,RA
     13,RAWR4,RES1,RES2,RES3,RES4
      DO 8921 I=N1A,N2
      RAWR1=-PROB1(I)
      RAWR2=(1.0-PROB2(I))
      RAWR3=-PROB3(I)
      RAWR4=-PROB4(I)
      RES1=RAWR1/C10(I)
      RES2=RAWR2/C11(I)
      RES3=RAWR3/C12(I)
      RES4=RAWR4/C13(I)
8921  WRITE(6,8902) PROB1(I),PROB2(I),PROB3(I),PROB4(I),RAWR1,RAWR2,RA
     13,RAWR4,RES1,RES2,RES3,RES4
      DO 8922 I=N2A,N3
      RAWR1=-PROB1(I)
      RAWR2=-PROB2(I)
      RAWR3=(1.0-PROB3(I))
      RAWR4=-PROB4(I)
      RES1=RAWR1/C10(I)
      RES2=RAWR2/C11(I)
      RES3=RAWR3/C12(I)
      RES4=RAWR4/C13(I)
8922  WRITE(6,8902) PROB1(I),PROB2(I),PROB3(I),PROB4(I),RAWR1,RAWR2,R
     13,RAWR4,RES1,RES2,RES3,RES4
      DO 8923 I=N3A,NSAMPL
      RAWR1=-PROB1(I)
      RAWR2=-PROB2(I)
      RAWR3=-PROB3(I)
      RAWR4=(1.0-PROB4(I))
      RES1=RAWR1/C10(I)
      RES2=RAWR2/C11(I)
      RES3=RAWR3/C12(I)
      RES4=RAWR4/C13(I)
8923  WRITE(6,8902) PROB1(I),PROB2(I),PROB3(I),PROB4(I),RAWR1,RAWR2,R
     13,RAWR4,RES1,RES2,RES3,RES4
8902  FORMAT(' ',12(F8.5,2X))
      DO 243 I=1,K
      PESTS1(I)=XZ1(I,1)
      PESTS2(I)=XZ2(I,1)
243   PESTS3(I)=XZ3(I,1)
      CALL PMAP(NL,NC,PESTS1,PESTS2,PESTS3,WORK,REFC,CINT,UAX,VAX,N,I
     1,VMAX,UMIN,UMAX,VMIN)
      RETURN
      END



      SUBROUTINE CALCFG(N,X,F,G)
      DOUBLE PRECISION X,G,U12,U13,U18,U23,F,DBLE
      COMMON/ONE/A(250,15),A2(100,15),A3(100,15),A4(100,15),A1(150,15
     1SAMPL,N1,N2,N3,N1A,N2A,N3A,N2B,N3B,K,K1,KA2,K2
      DIMENSION X(45),G(45),Q1(100,1),Q2(100,1),Q3(100,1)
      DIMENSION XN1(15,1),XN2(15,1),XN3(15,1)
      DIMENSION U6(250,1),U7(250,1),U8(250,1),U9(250,1)
      DIMENSION ROW1(1,250),ROW2(1,100),ROW3(1,100),ROW4(1,100)
```

```
      DIMENSION U11(250,15),U12(1,15),U13(1,15),U18(1,15),U23(1,15)
      J2=0
      J3=0
      DO 66 I=1,K
;6    XN1(I,1)=SNGL(X(I))
      DO 67 I=K1,KA2
      J2=J2+1
;7    XN2(J2,1)=SNGL(X(I))
      DO 68 I=K2,N
      J3=J3+1
 8    XN3(J3,1)=X(I)
      U1=0.0
      U2=0.0
      U3=0.0
      Z1=0.0
      CALL MC01AS(A2,XN1,Q1,N1,K,1,100,15,100)
      CALL MC01AS(A3,XN2,Q2,N2B,K,1,100,15,100)
      CALL MC01AS(A4,XN3,Q3,N3B,K,1,100,15,100)
      DO 169 I=1,N1
69    U1=U1+Q1(I,1)
      DO 70 I=1,N2B
0     U2=U2+Q2(I,1)
      DO 71 I=1,N3B
1     U3=U3+Q3(I,1)
      Z=U1+U2+U3
      CALL MC01AS(A,XN1,U6,NSAMPL,K,1,250,15,250)
      DO 87 I=1,NSAMPL
7     U6(I,1)=EXP(U6(I,1))
      CALL MC01AS(A,XN2,U7,NSAMPL,K,1,250,15,250)
      DO 88 I=1,NSAMPL
8     U7(I,1)=EXP(U7(I,1))
      CALL MC01AS(A,XN3,U8,NSAMPL,K,1,250,15,250)
      DO 89 I=1,NSAMPL
9     U8(I,1)=EXP(U8(I,1))
      DO 90 I=1,NSAMPL
      U9(I,1)=U6(I,1)+U7(I,1)+U8(I,1)
0     U9(I,1)=1.0+U9(I,1)
      DO 91 I=1,NSAMPL
1     Z1=Z1+ALOG(U9(I,1))
      FA=-1.0*(Z-Z1)
      F=DBLE(FA)
      ***** -F IS THE LOG LIKELIHOOD VALUE ****************************

      *** CALCULATE 1ST DERIVATIVES OF LOG LIKELIHOOD FUNCTION ********
      DO 44 I=1,NSAMPL
      ROW1(1,I)=1.0
      DO 45 I=1,N1
      ROW2(1,I)=1.0
      DO 46 I=1,N2B
      ROW3(1,I)=1.0
      DO 47 I=1,N3B
      ROW4(1,I)=1.0
      DO 94 I=1,NSAMPL
      U6(I,1)=U6(I,1)/U9(I,1)
      DO 92 J=1,K
      DO 92 I=1,NSAMPL
      U11(I,J)=A(I,J)*U6(I,1)
      CALL MATNW(ROW1,U11,U12,1,NSAMPL,K,1,250,1)
      CALL MATNW(ROW2,A2,U13,1,N1,K,1,100,1)
      DO 93 J=1,K
      U13(1,J)=U13(1,J)-U12(1,J)
      DO 95 I=1,NSAMPL
      U7(I,1)=U7(I,1)/U9(I,1)
      DO 96 J=1,K
      DO 96 I=1,NSAMPL
      U11(I,J)=A(I,J)*U7(I,1)
      CALL MATNW(ROW1,U11,U12,1,NSAMPL,K,1,250,1)
      CALL MATNW(ROW3,A3,U18,1,N2B,K,1,100,1)
      DO 97 J=1,K
      U18(1,J)=U18(1,J)-U12(1,J)
      DO 98 I=1,NSAMPL
      U8(I,1)=U8(I,1)/U9(I,1)
      DO 99 J=1,K
      DO 99 I=1,NSAMPL
      U11(I,J)=A(I,J)*U8(I,1)
      CALL MATNW(ROW1,U11,U12,1,NSAMPL,K,1,250,1)
      CALL MATNW(ROW4,A4,U23,1,N3B,K,1,100,1)
      DO 100 J=1,K
      U23(1,J)=U23(1,J)-U12(1,J)
      DO 101 I=1,K
      G(I)=U13(1,I)
      DO 102 I=1,K
      J4=I+K
```

```
  102 G(J4)=U18(1,I)
      DO 103 I=1,K
      J5=I+KA2
  103 G(J5)=U23(1,I)
      DO 104 I=1,N
  104 G(I)=-G(I)
C     ****  G IS THE VECTOR OF 1ST DERIVATIVES ************************
      RETURN
      END




      SUBROUTINE PMAP(NL,NC,PESTS1,PESTS2,PESTS3,WORK,REFC,CINT,UAX,V
     1N,IORD,VMAX,UMIN,UMAX,VMIN)
      DIMENSION PESTS1(N),PESTS2(N),PESTS3(N)
      DIMENSION WORK(N),SLINE(120),SYMB(23)
      DATA SYMB/1H1,1H.,1H2,1H.,1H3,1H.,1H4,1H.,1H5,1H.,1H\,1H.,1H6,1
     11H7,1H.,1H8,1H.,1H9,1H.,1H0,1HI,1H-/
    1 FORMAT('1         PROBABILITY SURFACE OF ORDER ',I8/' PART',I4//)
    2 FORMAT(' ',120A1)
      DO 66 JZ=1,4
      KMAX=0
      KMIN=20
      NO2=110
      NSTR=NC/110
      LEFT=NC-110*NSTR
      IF(LEFT.GT.0) NSTR=NSTR+1
      DO 96 KOUNT=1,NSTR
      IF(KOUNT.EQ.NSTR.AND.LEFT.GT.0) NO2=LEFT
      WRITE(6,1) IORD,KOUNT
      NO22=NO2+2
      WRITE(6,2) (SYMB(23),I=1,NO22)
      V=VMAX+VAX
      DO 12 L=1,NL
      V=V-VAX
      U=UMIN-UAX+(KOUNT-1)*110*UAX
      DO 6 I=1,NO2
      U=U+UAX
      PR=PROB(PESTS1,PESTS2,PESTS3,WORK,N,IORD,U,V,JZ)
      KCC=11+INT((PR-REFC)/CINT)
      IF(PR.LT.REFC) KCC=KCC-1
      IF(KCC.LT.1) KCC=1
      IF(KCC.GT.20) KCC=20
      IF(KCC.GT.KMAX) KMAX=KCC
      IF(KCC.LT.KMIN) KMIN=KCC
    6 SLINE(I)=SYMB(KCC)
   12 WRITE(6,2) SYMB(22),(SLINE(I),I=1,NO2),SYMB(22)
      WRITE(6,2) (SYMB(23),I=1,NO22)
   96 CONTINUE
      WRITE(6,4) REFC,CINT
    4 FORMAT('0     REFERENCE CONTOUR = ',F11.4/'      CONTOUR INTERVA
     1',F12.4)
      WRITE(6,3)
    3 FORMAT('0 KEY TO CONTOUR VALUES'//)
      DO 5 I=KMIN,KMAX
      J=11-I
      P=REFC-J*CINT
      Q=P+CINT
      IF(I.EQ.1) GO TO 7
      IF(I.EQ.20) GO TO 8
      WRITE(6,9) (SYMB(I),J=1,6),P,Q
      GO TO 5
    7 WRITE(6,10) (SYMB(I),J=1,6),P
   10 FORMAT(' ',6A1,2X,F10.4,'AND BELOW')
      GO TO 5
    8 WRITE(6,11) (SYMB(I),J=1,6),P
   11 FORMAT(' ',6A1,2X,F10.4,'AND ABOVE')
    9 FORMAT(' ',6A1,2X,F10.4,' TO ',F10.4)
    5 CONTINUE
      WRITE(6,13) UMIN,VMAX,UMAX,VMAX,UMIN,VMIN,UMAX,VMIN
   13 FORMAT('0 COORDINATES OF MAP CORNERS ARE:'/
     15X,' TOP LEFT',5X,2F12.4/
     15X,' TOP RIGHT    ',2F12.4/
     15X,' BOTTOM LEFT  ',2F12.4/
     15X,' BOTTOM RIGHT ',2F12.4)
   66 CONTINUE
      RETURN
      END
```

```
      FUNCTION PROB(PESTS1,PESTS2,PESTS3,WORK,N,NORD,U,V,JZ)
      *** COMPUTES VALUE ON PROBABILITY SURFACE AT EACH POINT *********
      DIMENSION PESTS1(N),PESTS2(N),PESTS3(N),WORK(N)
      QZ1A=PESTS1(1)+(PESTS1(2)*U)+(PESTS1(3)*V)
      QZ1B=PESTS2(1)+(PESTS2(2)*U)+(PESTS2(3)*V)
      QZ1C=PESTS3(1)+(PESTS3(2)*U)+(PESTS3(3)*V)
      QZ2A=EXP(QZ1A)
      QZ2B=EXP(QZ1B)
      QZ2C=EXP(QZ1C)
      ZZ=1.0+QZ2A+QZ2B+QZ2C
      IF(JZ.EQ.2) GO TO 8
      IF(JZ.EQ.3) GO TO 9
      IF(JZ.EQ.4) GO TO 98
      PROB=QZ2A/ZZ
      GO TO 10
      PROB=QZ2B/ZZ
      GO TO 10
      PROB=QZ2C/ZZ
      GO TO 10
8     PROB=1.0/ZZ
0     CONTINUE
      IF(NORD.LT.2) RETURN
      WORK(2)=U
      WORK(3)=V
      DO 1 M=2,NORD
      L1=M-1
      L2=M*(M+1)/2
      DO 2 I=1,M
      L2=L2+1
      NW200=L2-M
      WORK(L2)=WORK(NW200)*U
      QZ1A=QZ1A+PESTS1(L2)*WORK(L2)
      QZ1B=QZ1B+PESTS2(L2)*WORK(L2)
      QZ1C=QZ1C+PESTS3(L2)*WORK(L2)
      NW201=L2+1
      NW202=L2-M
      WORK(NW201)=WORK(NW202)*V
      QZ1A=QZ1A+PESTS1(NW201)*WORK(NW201)
      QZ1B=QZ1B+PESTS2(NW201)*WORK(NW201)
      QZ1C=QZ1C+PESTS3(NW201)*WORK(NW201)
      QZ2A=EXP(QZ1A)
      QZ2B=EXP(QZ1B)
      QZ2C=EXP(QZ1C)
      ZZ=1.0+QZ2A+QZ2B+QZ2C
      IF(JZ.EQ.2) GO TO 4
      IF(JZ.EQ.3) GO TO 5
      IF(JZ.EQ.4) GO TO 99
      PROB=QZ2A/ZZ
      GO TO 1
      PROB=QZ2B/ZZ
      GO TO 1
      PROB=QZ2C/ZZ
      GO TO 1
      PROB=1.0/ZZ
      CONTINUE
      RETURN
      END
```

APPENDIX 3 Derivation of the standard errors and variance-covariance matrix of parameter estimates

When the method of maximum likelihood is used to compute the parameter estimates, the asymptotic variance-covariance matrix is given by the inverse of the so called 'information matrix', $\underline{I}(\underline{\beta})$, (see Edwards, 1972; Cox, 1970, p.87) where

$$\underline{I}(\underline{\beta}) = -E\left[\frac{\partial^2 \log_e \Lambda}{\partial \underline{\beta} \partial \underline{\beta}'}\right]$$

and is consistently estimated by $\underline{I}(\hat{\underline{\beta}})$. $\underline{\beta}$ in this case is a vector of parameters, e.g. $\underline{\beta}' = [\alpha, \beta_1, \beta_2 \ldots]$. The programs listed in Tables 6, 7 and 8 print both the asymptotic variance-covariance matrix of parameter estimates and the asymptotic standard errors found by taking square roots of the diagonal of $\underline{I}^{-1}(\hat{\underline{\beta}})$.

able 9

```
  SUBROUTINE VA06AD(N,X,F,G,STEP,ACC,MAXFUN,IPRINT,W)
  DOUBLE PRECISION ACCT,ACC,DSS,STEP,X,F,G,GSQ,GGDIAG,HDIAG,W,DGGD,S
 1GDD,SUM,DSQ,C,GGGG,CB,CA,CC,THETA,S,DG,DGA,WBSQ,FA,SDM,GHD,SEG,DHD
 1,HDIV,CD,CE
  DIMENSION X(1),G(1),W(1)
  THE NEXT SEVEN INTEGERS PARTITION THE ARRAY W
  IDD=N*N+N
  IH=IDD/2
  IXA=IDD+N*N
  IGA=IXA+N
  IWA=IGA+N
  IWB=IWA+N
  IWC=IWB+N
  SET SOME CONSTANTS
  ACCT=ACC*ACC
  IPP=IPRINT*IPRINT
  GIVE INITIAL VALUES TO SOME VARIABLES
  DSS=STEP*STEP
  MAXC=1
  ITSPEC=1
  IPTEST=1
  CALCULATE THE INITIAL GRADIENT
  CALL CALCFG (N,X,F,G)
  GIVE INITIAL VALUES TO THE COMPONENTS OF GG, H AND DD
  GSQ=0.D0
  DO 1 I=1,N
1 GSQ=GSQ+G(I)**2
  IF (GSQ) 5,5,2
2 GGDIAG=0.01D0*DSQRT(GSQ)/STEP
  HDIAG=1.D0/GGDIAG
  K=0
  KDD=IDD
  DO 4 I=1,N
  J=1
1 IF (J-I) 102,103,103
2 KDD=KDD+1
  W(KDD)=0.D0
  J=J+1
  GO TO 101
3 K=K+1
  KDD=KDD+1
  W(K)=GGDIAG
  NW1=IH+K
  W(NW1)=HDIAG
  W(KDD)=1.D0
3 J=J+1
  IF (J-N) 104,104,4
4 K=K+1
  KDD=KDD+1
  W(K)=0.D0
  NW2=IH+K
  W(NW2)=0.D0
  W(KDD)=0.D0
  GO TO 3
4 CONTINUE
5 IF (IPP) 10,10,6
6 PRINT 7
  FORMAT(///,1X,'THE FOLLOWING OUTPUT IS PROVIDED BY VA06AD')
  GO TO 10
  BEGIN AN ITERATION BY TESTING FOR CONVERGENCE
8 GSQ=0.D0
  DO 9 I=1,N
9 GSQ=GSQ+G(I)**2
0 IF (GSQ-ACCT) 11,11,18
  PRINT THE FINAL VALUES OF THE FUNCTION AND GRADIENT
1 IF (IPP) 17,17,12
2 PRINT 13,MAXC,F
3 FORMAT(/5X,'AFTER',I4,' CALLS OF CALCFG, THE FINAL F =',
 1D14.6)
  PRINT 14,(X(I),I=1,N)
4 FORMAT(5X,'X =',(8D14.6))
  IF (IPRINT) 17,17,15
5 PRINT 16,(G(I),I=1,N)
6 FORMAT(5X,'G =',(8D14.6))
7 RETURN
  TEST WHETHER MAXFUN CALLS OF CALCFG HAVE BEEN MADE
8 IF (MAXC-MAXFUN) 21,19,19
9 PRINT 20,MAXC
```

```
   20 FORMAT(/5X,'VA06AD HAS MADE',I5,' CALLS OF CALCFG')
      GO TO 11
C     PRINT THE CURRENT BEST VALUE OF F ETC
   21 IPTEST=IPTEST-IABS(IPRINT)
      IF (IPTEST) 22,22,25
   22 IPTEST=IPP
      PRINT 23,MAXC,F
   23 FORMAT(/5X,'AT THE START OF ITERATION',I4,5X,'F =',D14.6)
      PRINT 14,(X(I),I=1,N)
      IF (IPRINT) 25,25,24
   24 PRINT 16,(G(I),I=1,N)
C     TEST WHETHER A SPECIAL ITERATION IS NEEDED, AND CALCULATE
C     THE CHANGE IN GRADIENT ALONG THE DIRECTION OF
C     A SPECIAL ITERATION
   25 ITSPEC=ITSPEC-1
      IF(ITSPEC)26,32,32
   26 DGGD=0.D0
      SGDD=0.D0
      KDD=IDD
      DO 28 I=1,N
      SUM=0.D0
      K=I
      KD=IDD
      J=1
  105 IF (J-I) 106,107,107
  106 KD=KD+1
      SUM=SUM+W(K)*W(KD)
      K=K+N-J
      J=J+1
      GO TO 105
  107 DO 27 J=I,N
      KD=KD+1
      SUM=SUM+W(K)*W(KD)
   27 K=K+1
      KDD=KDD+1
      SGDD=SGDD+G(I)*W(KDD)
   28 DGGD=DGGD+SUM*SUM
C     CALCULATE THE CORRECTION FOR A SPECIAL ITERATION
C     AND REVISE THE ARRAY DD
      DSQ=DMIN1(DSS,GSQ/DGGD)
      C=DSIGN(DSQRT(DSQ),-SGDD)
      DO 29 I=1,N
      NW3=I+IDD
      NW4=I+IWA
      NW5=I+IWB
      W(NW4)=C*W(NW3)
 29   W(NW5)=W(NW3)
      KDD=IDD
      DO 30 I=2,N
      DO 30 J=1,N
      KDD=KDD+1
      NW6=KDD+N
 30   W(KDD)=W(NW6)
      DO 31 I=1,N
      KDD=KDD+1
      NW7=I+IWB
 31   W(KDD)=W(NW7)
      ITSPEC=2
      GO TO 51
C     CALCULATE THE GENERALIZED NEWTON CORRECTION TO X
C     AND PREDICT THE CURVATURE OF F ALONG THE GRADIENT
   32 GGGG=0.D0
      DO 34 I=1,N
      NW8=I+IWA
      W(NW8)=0.D0
      SUM=0.D0
      J=1
      K=I
  108 IF (J-I) 109,110,110
 109  NW9=IH+K
      W(NW8)=W(NW8)-W(NW9)*G(J)
      SUM=SUM+W(K)*G(J)
      K=K+N-J
      J=J+1
      GO TO 108
  110 DO 33 J=I,N
      NW11=IH+K
      W(NW8)=W(NW8)-W(NW11)*G(J)
      SUM=SUM+W(K)*G(J)
   33 K=K+1
   34 GGGG=GGGG+SUM*G(I)
```

```
C     TEST WHETHER TO SET THE CORRECTION TO A MULTIPLE OF
C     THE GRADIENT
      IF (GGGG*DABS(GGGG)*DSS-GSQ**3) 35,35,37
C     SET THE CORRECTION VECTOR TO A MULTIPLE OF THE GRADIENT
   35 C=-DSQRT(DSS/GSQ)
      DO 36 I=1,N
      NW12=I+IWA
   36 W(NW12)=C*G(I)
      GO TO 41
C     SET THE OPTIMAL STEEPEST DESCENT CORRECTION IN WB
C     AND THE DIFFERENCE BETWEEN WA AND WB IN WC
   37 C=-GSQ/GGGG
      CB=0.D0
      DO 38 I=1,N
      NW13=I+IWB
      NW14=I+IWA
      NW15=I+IWC
      W(NW13)=C*G(I)
      W(NW15)=W(NW14)-W(NW13)
   38 CB=CB+W(NW15)**2
      IF (CB) 35,35,85
C     CALCULATE THE SCALAR PRODUCT (-W,GW)
   85 CA=0.D0
      DO 90 I=1,N
      CC=0.D0
      J=1
      K=I
   86 IF (J-I) 87,88,88
   87 NW16=J+IWC
      CC=CC+W(K)*W(NW16)
      K=K+N-J
      J=J+1
      GO TO 86
   88 DO 89 J=I,N
      NW17=J+IWC
      CC=CC+W(K)*W(NW17)
   89 K=K+1
      NW18=I+IWC
   90 CA=CA-CC*W(NW18)
C     INTERPOLATE FOR THE CORRECTION VECTOR ON THE LINE WA - WB
      CA=C*CA
      C=DSS-C*C*GSQ
      THETA=DSIGN(C/(DABS(CA)+DSQRT(CA*CA+C*CB)),CA)
C     TEST WHETHER TO USE THE GENERALIZED NEWTON CORRECTION
      IF (THETA-1.D0) 39,41,41
   39 DO 40 I=1,N
      NW19=I+IWA
      NW20=I+IWB
      NW21=I+IWC
   40 W(NW19)=W(NW20)+THETA*W(NW21)
C     EXPRESS THE CORRECTION VECTOR IN TERMS OF THE ROWS OF DD
   41 DSQ=0.D0
      KDD=IDD
      DO 42 I=1,N
      NW22=I+IWA
      NW23=I+IWB
      NW24=I+IWC
      DSQ=DSQ+W(NW22)**2
      W(NW23)=0.D0
      W(NW24)=0.D0
      DO 42 J=1,N
      KDD=KDD+1
      NW26=IWA+J
   42 W(NW23)=W(NW23)+W(KDD)*W(NW26)
C     REVISE THE DIRECTIONS IN THE ARRAY DD
      S=0.D0
      KK=N
   43 NW27=KK+IWB
      IF (W(NW27)) 45,44,45
   44 KK=KK-1
      GO TO 43
   45 KK=KK-1
      IF (KK) 48,48,46
   46 NW28=IWB+KK+1
      S=S+W(NW28)**2
      NW29=IWB+KK
      C=DSQRT(S*(S+W(NW29)**2))
      CA=S/C
      CB=W(NW29)/C
      KDD=IDD+N*KK
      DO 47 J=1,N
```

```
      KDD=KDD+1
      NW30=J+IWC
      NW31=IWB+KK+1
      W(NW30)=W(NW30)+W(NW31)*W(KDD)
      NW32=KDD-N
47    W(KDD)=CA*W(NW32)-CB*W(NW30)
      GO TO 45
   48 KDD=IDD
      DO 49 I=2,N
      DO 49 J=1,N
      KDD=KDD+1
      NW33=KDD+N
49    W(KDD)=W(NW33)
      C=1.D0/DSQRT(DSQ)
      DO 50 I=1,N
      KDD=KDD+1
      NW34=I+IWA
 50   W(KDD)=C*W(NW34)
C     APPLY THE CORRECTION VECTOR AND CHECK WHETHER
C     ROUND OFF MAKES THE CORRECTION ZERO
   51 DSQ=0.D0
      DO 52 I=1,N
      NW35=I+IXA
      NW36=I+IWA
      W(NW35)=X(I)+W(NW36)
      W(NW36)=W(NW35)-X(I)
  52  DSQ=DSQ+W(NW36)**2
      IF (DSQ) 80,80,82
   80 PRINT 81
   81 FORMAT (/5X,'ERROR EXIT FROM VA06AD')
      GO TO 11
C     CALCULATE THE NEXT VALUE OF THE OBJECTIVE FUNCTION
   82 MAXC=MAXC+1
C     NOTE THAT THE NEXT INSTRUCTION IS NOT STANDARD FORTRAN
      NW100=IXA+1
      NW101=IGA+1
      CALL CALCFG(N,W(NW100),FA,W(NW101))
C     SET THE ERROR OF THE PREDICTED GRADIENT IN WB
C     ALSO CALCULATE SOME NUMBERS FOR REVISING THE STEP-BOUND
      DG=0.D0
      DGA=0.D0
      DGGD=0.D0
      WBSQ=0.D0
      DO 54 I=1,N
      SUM=0.D0
      J=1
      K=I
  111 IF (J-I) 112,113,113
 112  NW37=J+IWA
      SUM=SUM+W(K)*W(NW37)
      K=K+N-J
      J=J+1
      GO TO 111
  113 DO 53 J=I,N
      NW38=J+IWA
      SUM=SUM+W(K)*W(NW38)
   53 K=K+1
      NW39=I+IWB
      NW40=I+IGA
      NW41=I+IWA
      W(NW39)=W(NW40)-G(I)-SUM
      DG=DG+G(I)*W(NW41)
      DGA=DGA+W(NW40)*W(NW41)
      DGGD=DGGD+SUM*W(NW41)
 54   WBSQ=WBSQ+W(NW39)**2
      IF (ITSPEC-2) 55,60,60
C     CHECK WHETHER ROUND OFF ERRORS ARE SERIOUS
   55 C=DG+.5D0*DGGD
      IF (C) 83,80,80
   83 CA=F+C
      CB=CA-F
      IF((CB-.5D0*C)*(CB-1.5D0*C))84,80,80
C     TEST WHETHER TO DECREASE THE STEP BOUND
   84 IF(FA-F-.1D0*C)57,57,56
   56 DSS=0.25D0*DSQ
      GO TO 60
C     TEST WHETHER TO INCREASE THE STEP-BOUND
   57 DSS=DSQ
      IF (WBSQ-0.25D0*GSQ) 59,59,58
   58 IF (DG-DGA-DGA) 60,59,59
   59 DSS=4.D0*DSQ
```

```
C     SET THE DIFFERENCE BETWEEN GRADIENTS
   60 DO 61 I=1,N
      NW42=I+IWC
      NW43=I+IGA
  61  W(NW42)=W(NW43)-G(I)
C     SET X, F AND G TO THE BEST CALCULATED VALUES
      ITHETA=1
      IF (F-FA) 64,64,62
   62 F=FA
      DO 63 I=1,N
      NW44=I+IXA
      NW45=I+IGA
      X(I)=W(NW44)
 63   G(I)=W(NW45)
C     CALCULATE SOME VECTORS AND SCALAR PRODUCTS TO
C     REVISE GG AND H
   64 SDM=0.D0
      GHD=0.D0
      SEG=0.D0
      DHD=0.D0
      DO 66 I=1,N
      NW46=I+IXA
      NW47=I+IWA
      NW48=I+IGA
      W(NW46)=-W(NW47)
      W(NW48)=0.D0
      K=IH+I
      J=1
  114 IF (J-I) 115,116,116
 115  NW50=J+IWC
      NW52=J+IWA
      W(NW46)=W(NW46)+W(K)*W(NW50)
      W(NW48)=W(NW48)+W(K)*W(NW52)
      K=K+N-J
      J=J+1
      GO TO 114
  116 DO 65 J=I,N
      NW54=J+IWC
      NW56=J+IWA
      W(NW46)=W(NW46)+W(K)*W(NW54)
      W(NW48)=W(NW48)+W(K)*W(NW56)
   65 K=K+1
      NW58=I+IWB
      NW59=I+IWC
      SDM=SDM+W(NW47)*W(NW58)
      GHD=GHD+W(NW59)*W(NW48)
      SEG=SEG+W(NW59)*W(NW46)
 66   DHD=DHD+W(NW47)*W(NW48)
C     TEST WHETHER THE USUAL CORRECTION TO GG GIVES
C     NEAR SINGULARITY
      HDIV=SEG*DHD-GHD*GHD
      GO TO (67,70),ITHETA
   67 IF(DABS(HDIV)-0.1D0*DSQ*DSQ) 68,70,70
C     CHANGE THE DIFFERENCE IN GRADIENTS TO AVOID SINGULARITY
   68 CA=HDIV/(DSQ*DSQ)+0.1D0
      CB=GHD/DSQ-0.1D0
      CA=CA/(CA+CB+DSIGN(DSQRT(0.9D0*CA+CB*CB),CA+CB))
      CB=(1.D0-CA)*SDM/DSQ
      DO 69 I=1,N
      NW60=I+IWA
      NW61=I+IWB
      NW62=I+IWC
      C=CA*(CB*W(NW60)-W(NW61))
      W(NW61)=W(NW61)+C
 69   W(NW62)=W(NW62)+C
      ITHETA=2
      GO TO 64
C     REVISE THE MATRICES GG AND H
   70 CA=1.D0/DSQ
      CB=SDM*CA*CA
      CC=DHD/HDIV
      CD=GHD/HDIV
      CE=SEG/HDIV
      K=0
      DO 71 I=1,N
      DO 71 J=I,N
      K=K+1
      NW63=I+IWA
      NW64=J+IWB
      NW65=J+IWA
      NW66=I+IWB
      NW67=K+IH
```

```
      NW68=I+IXA
      NW69=J+IXA
      NW70=I+IGA
      NW71=J+IGA
      W(K)=W(K)+CA*(W(NW63)*W(NW64)+W(NW65)*W(NW66))-CB*W(NW63)*W(NW6
71    W(NW67)=W(NW67)-CC*W(NW68)*W(NW69)-CE*W(NW70)*W(NW71)+CD*(W(NW6
     1W(NW71)+W(NW69)*W(NW70))
      GO TO 8
      END
```

```
      SUBROUTINE MC01AS(A,B,C,L,M,N,IDA,IDB,IDC)
      DOUBLE PRECISION DA,DB,SUM,DBLE
      DIMENSION A(IDA,IDB),B(IDB,N),C(IDC,N)
      L1=L
      M1=M
      N1=N
      DO 1 I=1,L1
      DO 1 J=1,N1
      SUM=0.0D0
      DO 2 K=1,M1
      DA=DBLE(A(I,K))
      DB=DBLE(B(K,J))
2     SUM=SUM+DA*DB
1     C(I,J)=SNGL(SUM)
      RETURN
      END
```

```
      SUBROUTINE MATNW(V1,V2,V3,IV1,IV2,IV3,IV4,IV5,IV6)
      DOUBLE PRECISION DV1,DV2,V3,VSUM,DBLE
      DIMENSION V1(IV4,IV5),V2(IV5,IV3),V3(IV6,IV3)
      LV1=IV1
      MV1=IV2
      NV1=IV3
      DO 1 I=1,LV1
      DO 1 J=1,NV1
      VSUM=0.0D0
      DO 2 K=1,MV1
      DV1=DBLE(V1(I,K))
      DV2=DBLE(V2(K,J))
2     VSUM=VSUM+DV1*DV2
1     V3(I,J)=VSUM
      RETURN
      END
```